BIRDS
in Fact
& Legend

BIRDS in Fact & Legend

Walter Harter

Sterling Publishing Co., Inc. New York
Oak Tree Press Co., Ltd.
London & Sydney

OTHER BOOKS OF INTEREST

Bird Is Born
Cage Bird Identifier
Fabulous Birds You Can Make
How to Raise & Train Pigeons
Pigeon Racing

Two Park Avenue, New York, N.Y. 10016
Distributed in Australia by Oak Tree Press Co., Ltd.,
P.O. Box J34, Brickfield Hill, Sydney 2000, N.S.W.
Distributed in the United Kingdom by Ward Lock Ltd.
116 Baker Street, London W.1
Manufactured in the United States of America

Library of Congress Catalog Card No.: 79-65065
Sterling ISBN 0-8069-3740-8 Trade Oak Tree 7061-2661-0
3741-6 Library

Contents

Introduction

There are nearly 9,000 kinds of birds, of all hues, sizes and characteristics. Some, like the hummingbird, are so small they can hardly be seen. One, the ostrich, is taller than a man.

Some birds—the pheasant, peacock, bird of paradise, and many members of the parrot family—are so beautifully pigmented that their feathers appear to have been painted by an inspired artist. While others, like the mudhen, limpkin and sparrow, are drab and plain.

Many birds sing lovely songs; others emit only raucous squawks. A few, like the hummingbird, make only faint sounds with their vocal organs.

Since early man glanced at the sky and became aware of the amazing world of feathers, birds have had an important part in his life. Not only have they filled his homes and lands with beauty and song, but they have helped him in many other ways.

Some have provided him with food, others have protected him, a few have been so important in his civilization that they became the emblems of nations.

Even before there were written languages, birds were part of man's folklore, superstitions and even religions. Birds have been, and are, representatives of hope and despair, of good luck and bad.

Much of the fascination of birds comes from the mystery that surrounds them, owing to the speed of their flight, the amazing keenness of their eyesight, and their ability to migrate over long distances, often at night, to places that are mere pin points on the map of the world.

And of course there are more legends about birds than there are feathers on the body of an eagle. Is a bald eagle bald? Why did Noah select a dove to dispatch from the Ark when the great flood began to subside? Why are ravens black when, according to legend, they once were white? Why do penguins "fly" underwater and not on land? How and why do some pigeons find their way home through darkness and rain?

This book will answer many of the questions and explore some of the mysteries that have surrounded these marvellous creatures for thousands of years.

How and Why Birds Fly

Birds have been called "feathered serpents" or "glorified reptiles," because they evolved from a branch of the dinosaurs during the Jurassic geological period, about 140,000,000 years ago.

But much of the similarity disappeared long ago. Although the reptiles have stayed much as they were, birds gradually changed. Scales became feathers, and forelimbs changed into wings. Birds developed a circulatory system that maintains a constant body temperature; like mammals, they became "warm-blooded."

One of the most important changes was in their bones. These slowly became hollow, almost weightless, allowing birds to soar on the winds. Also feathers, acting as insulation, enabled birds to live and fly in all parts of the world, even in the polar regions. Their mouths gradually became toothless, protected by a hard bill. The only scales remaining from their reptilian origin are on their feet.

Over the many centuries birds' bodies underwent more changes. The wings became merely bony frames covered with feathers. And, as with our modern super-planes, the bird's center of gravity, its body, became the muscle-power source. From there a complicated system of thin sinews forces the lightweight wings to beat, without causing imbalance. Also, intricate air-sacs have evolved to store the huge amounts of oxygen required for swift and long flights. Another important evolution permits birds' eyes to adapt instantly when they swoop from a glittering sunlit sky into the shade of a forest.

For thousands of years man has tried to imitate the flight of a bird. Leonardo da Vinci drew plans for flight-machines, and the Wright brothers finally put some of those ideas, plus their own, into reality at Kitty Hawk, North Carolina, in 1903. But nothing man has been able to create matches exactly the flying structure of a bird's body. Its ability to turn instantly, its swiftness, have never been equalled. Perhaps if there *are* flying saucers it's because beings in another galaxy have solved the secret of a bird's flight, and have manufactured machines to imitate it.

The way in which birds learned to fly could have come about in many ways. Maybe they climbed trees at first seeking insects and other food. Then, as their wings gradually grew larger, they might have fallen or leaped from high branches and glided clumsily to the ground or to another tree. Many more years would pass before they learned to beat their wings and really fly. The only other animals alive today that possess the power of true flight are the bats and many of the insects. All other so-called "flying" animals—flying fishes, flying squirrels and flying lemurs, for example—are gliders, not fliers, and none of them have true wings.

However it happened, we now can marvel at these astonishing creatures, thrill to their swift passage over our heads, and enjoy their beauty and sound. It's as if our world is a huge cake—and birds are its decorations and icing.

Preening

In most dictionaries one of the definitions of "preening" is "to dress, to adorn, to show satisfaction or vanity."

That's what a bird seems to be doing when it strokes its feathers with its bill, spreads its wings, and struts. All birds, large or small, do it, and their actions make many observers think the birds are trying to make themselves beautiful, or if they are already pretty, are trying to attract attention.

But that's not true. Vanity is only one of the human traits birds know nothing about. With birds, preening is a serious business that could be a matter of life or death to them. It's taking care of themselves, preparing for the everyday struggle for existence. It's as necessary to them as brushing our teeth and keeping ourselves clean are to us.

Water birds—pelicans, herons, and gulls, to name only a few—build up layers of nasty materials on their feathers. If not removed, this substance would slowly interfere with the use of their wings, and finally they would be unable to fly. The result would be starvation, or they could become easy prey for their enemies. Nature, being all wise, and with centuries of experience, has arranged matters so that birds can "renew" themselves.

Sea birds, being so often immersed in the slime and ooze along the coast, have self-cleaning aids built into some of their feathers. These very small feathers produce a powder-like substance that spreads to the larger plumes and absorbs the unwanted materials. Then the bird combs out the powder with its bill or feet.

However, many birds, especially those that live near water, are equipped with another natural aid to help them take care of their feathers. Beneath the tails of these birds is an oil-producing gland. This oil—a perfect water-proofing material—is spread on the feathers by the bill and feet. This application of oil aids the birds to slip in and out of the water easily, and also gives them more speed during flight. They must give themselves the treatment quite often, at least several times each week.

Frequently, this beauty treatment, as well as the combing and cleaning of the feathers, can't be done without help. That's why, very often, birds appear to be showing affection for one another, pecking, rubbing heads together, or combing each other's feathers. But usually this isn't love-making or courtship, it's simply putting into practice the old saying, "You scratch my back and I'll scratch yours."

However, some birds, those who live inland, have devised other ways to rid themselves of unwanted fleas and other parasites. Often they find anthills, and, by staying close to them, manage to entice the ants to leap onto their feathers and do a cleaning job. When the birds think they've had enough attention and are clean, they flutter their wings, shaking off the beauty operators, and fly away.

But no matter how the "renewing" is done, the cleaning and oiling of their feathers are the principal ways all birds prolong their survival.

Bird Architects

Some nests, or bird homes, rival in architecture many of our more elaborate and complex buildings. Birds *should* be more expert in constructing shelters for their families. They were doing it millions of years before man began placing one stone on top of another to erect protections from dangers and the elements.

Like the homes of humans, birds' nests range from shacks to palaces. Some birds, like the penguin, require only a pebble or two on which to lay their eggs and raise their young. Other birds create complicated apartment houses.

There are many reasons why birds' nests vary so dramatically. One, of course, is that birds are of many different sizes, and therefore require a diversity of accommodations. The nest of a hummingbird, which is constructed of fine plant material often held together with silk from spider webs, will weigh only a fraction of an ounce. An eagle's home may weigh up to a ton, and be 20 feet tall and 10 feet wide. Not only the size and weight of birds, but the number of their eggs, plus the availability of building materials, and climate, are responsible for the wide differences in bird homes.

The largest and most exotic of all nests is the one constructed by the Australian brush fowl. This bird, averaging only 20 inches in length, will, by using its beak and claws, erect a mound averaging 35 feet in diameter and 15 feet high, containing a series of tunnels in which the female deposits her eggs. This mound is actually a giant incubator. Each day, after the eggs have been laid, the male carefully examines the nest. By some strange means he determines if it is too warm or too cool. If it is too cool, he immediately adds more material (leaves and ground scrapings). If it is too warm, he tears away part of the structure. Every day this bird, weighing a mere 10 pounds, might move one ton of earth.

The most elaborate and attractive nests are made by the bowerbirds of New Guinea. Their homes are decorated with flowers and bits of paper, and are even painted—the birds apply crushed juices with twigs they use as "paint brushes."

Other strange nests are built by the swifts of almost every country. These birds use their own saliva as binding material. The small nests, usually cup-shaped, are attached to walls, the undersides of huge leaves, and even the trunks of trees. The saliva-glue is so strong that not one nest has ever been observed to have fallen by its own weight. This type of glue-architecture might be valuable to the Californians who build homes on hills—homes that tend to slip from their foundations during heavy rains.

The nests of certain swifts are the basis of the famous "bird's nest soup," so well known in China and the Far East. Although truckloads of the nests are gathered for consumption every year, the birds that build them are in no apparent danger of extinction.

Whether a nest is large or small, it is usually constructed

with care and precision. No matter how rough the outside of a nest might be, the interior is as comfortable as experience and available materials can make it: the principal reason for all nests is the rearing and protection of the young.

Some birds go to great lengths to make sure their progeny are comfortable. Most ducks and geese, for example, pluck the down from their own breasts to line their nests so their offspring will have the softest of beds. Even sparrows upholster their nests with their own feathers or any others that they discover on the ground.

And so, long before man sought shelter in caves, the birds were practicing rules of building that even now impress architects. Perhaps our homes might be sturdier and more comfortable if we studied the methods of the birds.

Free as a Bird

Don't believe it. The conduct of all birds is bound by as many rules and restrictions as are those that confine the rest of us poor creatures.

Watching birds flit from tree to tree, or seeing them soar so effortlessly and gracefully through the sky, it's logical to think they have absolute freedom—and so the expression, "free as a bird."

However, as ornithologists and amateur bird watchers know, birds are "prisoners" of the land they fly over and slaves to the air they fly through. In other words, they are subject to certain "rules" imposed by nature. Called "behavior patterns," these include certain things every species of birds "can't help doing."

Of course there are nonconformists in the bird world, just as there are among the human hordes that populate the land. A man might leave his wife, children and profession and begin another "life" completely different from his "pattern." Likewise a maverick bird might depart from the routine of his species and do exactly the opposite of what is expected. But these variants are rare. The majority of creatures, including humans, *conform* to rituals and patterns that have been established over many centuries.

Humans grow into adulthood, marry and create families. It's *normal* for the male to work and for the female to make a home and take care of the children. *Where* this is done has become unimportant because of modern communications and the ease of travel. (However, until very modern times it was *normal* for people to remain in the areas where they had been born.)

The strongest pattern followed by birds is a devotion to a certain piece of land. These "familiar" territories assure them of security, and therefore help perpetuate each species. Birds that are year-round residents of a territory guard it for life, and with their lives. Migrant birds have both summer and winter homes, and protect each one as fiercely as do birds that never roam.

Because many types of birds travel in flocks, they might appear to be trespassers over certain areas, but they are not. These flocks are really well-ordered "bird societies" made up of old and young. Bird residents of a certain area will allow flocks of a *different* species to cross their land, or even to stay for a few days. But flocks of the residents' species will be driven away. Groups of birds are often observed fighting in midair, with all the ferocity and manoeuvring of an air battle between human enemies. In these battles it's quite rare for a flock defending its "home territory" to be driven away by intruders.

Birds' territories are established by song. If a male bird sings (as he does almost continuously) and is not challenged by the song of another male, he knows the area is still his. But if another male of his species answers, then challenges him by flying at him, swooping and pecking, it's a warning that a newcomer wants the territory for *his* family, and is willing to fight for it. Usually, though, the challenger gives up quickly, and flies away to find some unoccupied area in which to set up housekeeping.

That is why, when a mirror is set up in a bird's territory, he will peck at it because, seeing his own reflection, he believes another of his own kind is trespassing on his property.

By occupying as much land as he can defend, a bachelor not only makes sure he'll have enough food, but also might acquire a female companion. An unattached female selects a mate first by his voice (birds' voices are as distinctive as are humans'), then by what type of territory he controls. Many bachelors who can find only small pieces of land on the edges of their flock's domain, will often have to remain loners until they manage to acquire larger and better estates.

Even on their own "turf," birds don't fly around aimlessly. They have definite routes to water and food. A bird will leave its nest by the same air "lane" every day, and return by the same route. In one experiment, posts were erected on one side of a feeding station. Birds who were accustomed to arrive from that direction actually flew into the obstacles, instead of changing their route.

"Free as a bird!" Not so. They, as do all creatures, live by conforming to instinctive and hereditary rules as demanding and confining as silken chains.

Feathers Anyone?

Ancient man used feathers to decorate his body, perhaps even before he wore furs. Cave pictures depict men and women in various activities, and all of them, especially the men, are bedecked with feathers. Most humans love adornment, and what could be more lovely than the brilliant plumage of the creatures that glide so mysteriously through the air?

Captain Cook, stepping ashore in Hawaii, was greeted by chiefs wearing long robes made of feathers, just as, across the world, Spanish explorers were astonished at the feathered finery of the Indians who welcomed them.

Farther north, the Indian chiefs wore feathered headdresses as they galloped their ponies across the plains. An Indian "brave" was allowed to stick one feather in his headband for every enemy he killed in battle. We say "a feather in his cap" when someone accomplishes an extraordinary feat. And, "showing the white feather," derives from the usually white tail feathers displayed by a vanquished bird as it retreats from a field of conflict.

Arrows were, and still are, tipped with feathers, as are the shuttlecocks we use in games. Nights in British pubs would be dull indeed without games played with darts having feathers on their ends. For, if a bird can slip so easily through the air, why can't its feathers help objects do the same?

Feather dusters still do a better job than any imitations. And many of our literary classics would not have been possible without the quill pen, made from a bird's stiff tail feathers.

What can induce sounder sleep than a pillow filled with goose feathers? Or when the body is warmed by an eiderdown quilt? Nothing made by man can equal the softness and warmth of the breast feathers of the eider duck. The most expensive sleeping bags and quilted coats we use today are stuffed with the now-rare breast "down" of the sea ducks of northern climates.

Originally, only human males decorated themselves with feathers, no doubt imitating the male birds who bore more brilliant plumage than did their drab mates. But everything changes, and eventually it was the human female who was more gorgeously adorned with the brilliant clothes of birds.

Beginning in the 16th century women's hats became sunbursts of feathers, often displaying entire stuffed birds. Ostrich feathers were among the most popular—160 tons of them were sold in France alone in 1912. The amount sold worldwide must have been staggering!

Not only hats, but fans and scarves—"boas"—were made from the graceful white feathers of the giant birds. Ostrich farms became profitable, for one flock will produce year after year. The tail feathers can be pulled from the creatures without harming them, and the birds promptly grow another crop.

Of course other feathers were used in huge amounts. One, the plumage of the egret, became literally worth its weight in gold!

Then, after an intense campaign to make people aware of the wanton carnage committed on wildlife simply to supply ornaments, laws were passed prohibiting dealing in feathers and plumes. The United States passed the first law in 1913; England followed in 1921. However, feathers are still used in some places as body decorations. In Las Vegas no show girl would consider herself properly clothed without at least one ostrich feather. And one of the most exotic dancers of all times, Sally Rand, would have been out of business without her ostrich fan.

Feathers are still used as status symbols, too. The Queen of England wears an ostrich feather in her hat when reviewing the troops. The Prince of Wales is entitled to wear three.

The Pigeon

There are 289 species of pigeons, ranging in size from some as small as a sparrow to others as large as a chicken. Some do tricks. "Rollers" fly to great heights, then "roll" over and over until they reach the ground. "Tumblers," when placed on the ground, turn somersaults again and again until they are stopped by some object. Some pigeons have grotesque beaks, others have long, trailing tail feathers. All are fascinating in individual ways.

But among the most amazing of the pigeons is one particular bird whose strange talent has helped win wars, saved lives, and contributed in many ways to man's safety, health and enjoyment.

This is the homing pigeon. In shape, size and pigmentation it might resemble the common pigeons we see perched on buildings and in trees. But comparing these to "homers" is like comparing plough horses to purebred Arabian steeds.

The strange talent possessed by the homing pigeon is its mysterious ability to fly directly to its nest—or cote—when it is released hundreds or thousands of miles from its home.

The breeding lines of these marvellous birds are as carefully preserved and guarded as those of the most costly racing horses. And some are just as expensive. Some homers, depending upon their breeding, sell for hundreds, even thousands of dollars.

The sport of pigeon racing is as old as history, for it was enjoyed by the Persians, Egyptians, Greeks and Romans. Through the centuries, kings as well as commoners have bred the birds for speed and endurance. In modern times two of the most famous breeders were Mussolini and Franklin Delano Roosevelt.

Some of President Roosevelt's birds carried messages to and from besieged Paris during World War II. And at the end of that conflict an interesting auction was held near Rome, when Mussolini's homing pigeons were sold to the highest bidders—pigeon fanciers who journeyed long distances to attend the event.

The sport has grown even more popular today. Not only in the United States, but in almost every country, there are pigeon racing-and-homing clubs. Races and tests are conducted several times every year, with prize money amounting to more than $100,000!

Feathered Messengers

Late in the afternoon of June 15, 1815, a mud-spattered coach, pulled by two lathered horses, left a rutted road outside a small village near the city of Liège, Belgium, and stopped in the shadow of some trees. Paul Julius Reuter, the man in the coach, had followed the French army for almost one hundred days, since Napoleon returned from exile on the island of Elba.

They had been strenuous days, but now Reuter was sure a climax was near. English and German armies had finally intercepted the French, and a furious battle was now being fought. From where he had stopped his coach Reuter could hear the sound of artillery, and could see the red and orange flashes on the horizon.

Reuter had a plan. He knew that the outcome of the battle would be of utmost importance to the destiny of all Europe. And that the news would be of immense value, in fame as well as money, to whoever could deliver it first. News of that importance was always sent by couriers on fast horses. But that method was not fast enough, and the dispatches were often captured, or delayed.

Beside Reuter in the coach were wooden cages containing a few pigeons. He had experimented with homing pigeons, and had brought these from his home at Calais, the French port on the English Channel.

During the long hours he waited, Reuter began to learn the result of the fierce battle taking place only a few miles away. Limping, wounded and crying French soldiers passed the coach, retreating towards Liège. He stopped a number of them, then made his decision. Napoleon was defeated.

Reuter scribbled the news on a piece of paper, enclosed it in a metal tube, and attached it to the leg of a pigeon. He repeated the action three times, hoping that one of the birds would arrive safely at Calais. All three flew directly to his pigeon loft. The news was immediately placed on board a fast boat and dispatched to London.

Those pigeons, flying from the battlefield at Waterloo, were the beginning of the famous Reuter's news agency. They were also the start of the Rothschild fortune. For Nathan Meyer Rothschild made excellent use of the information. He sold his French securities and bought German and British stocks before news of Napoleon's defeat reached other financial markets.

More than 160 years after that event two white-coated doctors stood at a window of the Devonport Hospital, Plymouth, England. They were attaching small glass capsules to the legs of some pigeons. In the capsules were samples of human blood and tissues. In a few minutes the doctors raised the window and gently tossed the birds into the air. Then, as they waited, one doctor stared at his watch.

When the telephone rang, he quickly answered it, then smiled at his companion. "All arrived safely." Then he added,with another glance at his watch, "Exactly four minutes."

For a few years important human samples had been sent to the Freedom Fields Laboratory, two miles away, for immediate analysis. But the automobiles used were too often delayed by traffic and breakdowns, and those delays cost lives. However, since homing pigeons have been used for the short journey there have been no delays, and not one capsule has been lost. Two miles is not a great distance, but those same pigeons would be capable of delivering the samples if the laboratory was 200 miles from the hospital.

Since long before the first Olympic Games, when homing pigeons sped the names of the winners to villages and towns in ancient Greece, messages have been sent by the fast-flying birds who head directly for their homes.

Homing pigeons have been used by the United States (as well as other countries) in wars, and there have been many bird heroes. The most famous was a pigeon named "Cher Ami," who, with one leg shot away and a wounded wing, managed to reach headquarters with information that saved an entire battalion. Cher Ami, with other pigeon heroes, is mounted and on display at the National Museum, Washington, D.C.

During wartime, homing pigeons were bred and trained by the Signal Corps, Fort Monmouth, New Jersey. When more birds were needed, private breeders gladly supplied them. Fortunately, homing pigeons continue to be bred and trained. For, as one official at the Pentagon said: "Who knows when they will be needed again? When electronic equipment is disabled, especially in planes, a homing pigeon will always deliver a call for help."

No doubt he was thinking of the hundreds of British and American airmen whose lives were saved when they released pigeons after being shot down over the English Channel.

The term "pigeonhole" comes from the round openings made beneath roofs of barns to allow the birds to enter and make nests. No doubt pigeonholes in desks often contain objects of worth. However, pigeonholes in roofs shelter things that sometimes are of greater value.

How *Do* Pigeons Find Their Way Home?

Why homing pigeons fly directly to their cotes can be answered easily. A pigeon doesn't fly for pay, or promotion, or because it has been ordered to—it flies home because that's where its mate and young are waiting.

If a homing pigeon, usually a male, is taken from its nest, then released, it will unhesitatingly find the most direct route home. The pigeons that flew with the names of the winners of the first Olympic Games, had first been taken from their cotes in the distant villages, just as the pigeons that delivered the news of Napoleon's defeat had been taken from their homes in Calais.

But *how* pigeons can make their way unerringly through strange areas, sometimes in darkness and rain, is what has intrigued scientists for centuries. Many theories have been put forward, and many experiments made, but even now the homing instinct of some pigeons remains a secret.

It was believed that the eyes had something to do with the mystery. They *are* extremely keen, as are the eyes of almost all birds. But a pigeon's eyes are quite different from others. They are controlled by six muscles, enabling the bird to see clearly in all directions. There is also a special muscle that permits each eye to concentrate its "seeing force" instantly on objects at short or long distances.

A pigeon's eye also has two eyelids: one like ours, and a transparent extra lid that falls over the eye when the bird is in flight, protecting it from wind and rain, exactly like the windshield of an automobile.

But, marvellous as the eyes are, they don't explain how a pigeon that has been confined in a closed basket, then transported in an automobile or plane for 500 miles—far beyond the horizon line—will set out for its home as soon as released.

Next the pigeon's ears were studied. Some scientists believed the canals in the ears developed a sensitive "memory" of the direction in which they were transported, and this memory directed the bird homeward. But that theory was disproved when pigeons were taken long distances in containers that rotated constantly. When released the birds didn't hesitate, but flew directly back to their nests.

There are theories that involve the rotation of the earth, and the effects of the sun on flying birds. But homing pigeons fly at night, too, and messages have been delivered in dense fog.

The latest theories revolve around the earth's magnetism, and its influence on a particular part of a pigeon's eye called the "pecten." Some now believe that the pecten is a kind of navigational instrument that makes use of the magnetic fields to determine direction.

In some experiments, when birds have been followed by planes, they do appear to fly a pattern similar to the mysterious magnetic fields. However, when magnets were attached to the pigeon's legs, erasing the effect of the fields, the birds flew home directly and swiftly.

Scientists in many of our universities are still experimenting with the mystery of the homing pigeon, hoping that solving the puzzle will be of some use in improving our electronic matériel. However, until now they have been unsuccessful.

As long as people of all lands continue to breed and train homing pigeons, these gentle birds will supply sport, help and entertainment. *How* they manage to find their way to their cotes from far distances is just one more of nature's mysteries man has been unable to solve.

The Birds of Peace and Love

Among the 289 species of pigeons there is one group that is very special. They are the doves. Throughout history these gentle, graceful and lovely birds have been symbols of peace and tranquility. While the term "dove" may be used interchangeably with "pigeon," it is usually applied to certain wild species that are smaller than the domestic pigeons.

Doves are similar to their pigeon cousins in many ways, even to producing in a gland at the base of their bills, a milk-like substance they feed their young. The voices of doves, consisting of *cooing* sounds, are softer than those of pigeons. A few types of doves are so gentle and have such soothing voices that they have been kept in cages from early history. Turtledoves are especially adapted to living with humans, and a pair of them makes a delightful addition to any room.

Some types of doves—the rock dove, white-fronted and white-winged doves—are more at home in the wild than near human habitations. However, one kind—the mourning dove—is most happy in gardens and around people. This dove, often mistakenly called the "morning" dove, has a pleasantly soft coo, delightful to hear, especially early in the day. These sounds are so greatly enjoyed by some people that they erect small houses, called dovecotes, on poles in their gardens or beneath the eaves of their houses to entice the birds to set up housekeeping.

The nests doves make for themselves are often flimsy affairs constructed of sticks and grass, and are placed almost anywhere —in trees, gutters of houses, along the roofs of barns, and even beneath bridges, but seldom on the ground. Usually, two whitish eggs are deposited each spring. After a few weeks of subsisting on the "milk" poured into their mouths by the females, the young join their parents in a diet of insects, fruit, grain and seeds.

In ancient Greece the dove became closely connected in many ways with gods and goddesses. Even Aphrodite (Venus), the goddess of love, was supposed to have come from an egg hatched by a dove, and then rolled ashore by a fish. That might have been responsible for the tale that Aphrodite "rose from the foam-tossed sea."

Even oracles were supposed to have come about as the result of the activity of doves. According to Bulfinch's *The Age of Fable,* the most ancient oracle, that of Zeus at Dodona, originated when two black doves took flight from Thebes in Egypt. One flew to Dodona in Epirus and, alighting in a grove of oaks, proclaimed in human language that the inhabitants should there establish an oracle. The other dove flew to the temple of Zeus in Libya and delivered a similar command at that place.

When Christianity became dominant in Italy, the dove was already regarded there as representative of all that was good. It even came to symbolize the third member of the godhead—the Holy Ghost, and still holds that place. In one of the most loved and often sung hymns are the words: "Come Holy Spirit, heavenly Dove."

Doves have been associated with Christianity from the time of the Crucifixion. In his *Birds of Ireland*, John Watters (Dublin, 1853), mentions:

> "It is said that a dove perched in the neighbourhood of the Holy Cross when the Redeemer was expiring and, wailing its notes of sorrow, kept repeating the words 'Kyrie! Kyrie!' (meaning 'Lord have mercy!') to alleviate the agony of His dying moments."

The lives of medieval saints and martyrs—or the records of them—abound in references to doves. Even many popes revered the bird. A dove is the emblem of Pope Gregory the Great (A.D. 590–604), and its figure rests on his right shoulder on the magnificent statue of that pope in Rome. This is told in *The Catholic Encyclopedia*:*

> "When the pope was dictating a sermon about Ezechiel a veil was drawn between his secretary and himself. When the pope remained silent for a long time, the secretary made a hole in the veil and, peering through it, saw a dove seated on Gregory's head with its beak between his lips. When the dove withdrew its beak the holy pontiff spoke and the secretary took down the words; but when he became silent the secretary again looked through the hole and saw that the dove had again placed its beak between his lips."

The dove is held in reverence in other religions, too. Humphrey Prideaux, the 18th-century English orientalist and dean

*Grolier Inc. New York.

of Norwich, in his *Life of Mahamet,* relates that the prophet taught a dove to sit on his shoulder and eat seed from his ear. The wily prophet "gave out that it was the Holy Ghost, in the form of a dove, come to impart to him the counsels of God." That's probably why Shakespeare, in *Henry V*, asks, "Was Mohammed inspired with a dove?"

Islam has preserved the Oriental reverence for the gentle birds. Flocks of them surround all the mosques. They are regarded as sacred, and never killed. There are newspaper accounts that, in 1921, two European boys killed a couple of doves and thereby provoked a riot in the streets of Bombay. The Moslems were horrified, and the police had difficulty suppressing the disturbance. The stock exchange and other businesses were closed, and a wide-spread strike of workmen in India was threatened, as evidence of the deep feeling aroused by the boys' sacrilegious act.

People in the Middle Ages believed there was something supernatural and even holy about doves. In Malory's *Morte d'Arthur* a dove, entering through a window of the castle, carried a gold censer in its beak, and impressed the awestruck knights of the Round Table as a token of purity. This incident is commemorated in the opera *Parsifal.* During medieval times another incident occurred that has become tradition, and is devoutly believed by Venetians—that the doves in the Piazza San Marco fly three times daily around the city in homage to the Trinity.

Perhaps the idea of the dove as a messenger and bringer of good tidings began with the bird Noah released from the Ark, and which returned carrying an olive leaf in its beak. However, during Hernando Cortez' first voyage to America, in 1504, the crew became discouraged and mutinous, and then, according to his record of the voyage:

". . . came a Dove flying to the shippe, being Good Friday

at Sunsett; and sat him on the Shippe-top; whereat they were all comforted, and tooke it for a miracle and good token . . . and all gave heartie thanks to God, directing our course the way the Dove flew."

The Bible often mentions doves as symbols of divine sanctions. Matthew, in chapter 3, verse 166, records "Lo, the heavens were opened unto him, and he saw the Spirit of God descending like a dove and lighting upon him." Luke mentions that "the Holy Ghost descended in a bodily shape like a dove."

There is little wonder that the dove has become so closely allied with Christ and Christianity, and that in many countries, especially those with a Catholic faith, the birds are revered.

The white dove has always been a symbol of purity, no doubt because its whiteness is a reminder of unstained snow and light. Perhaps for that reason white is always worn at confirmations and weddings.

At any rate the earth is a better place for having these lovely and gentle creatures as inhabitants. And what better alarm clock could a tense world have than the soft cooing of mourning doves?

The Feathered Pet of Kings and Queens

The tiny speck in the sky gracefully glided in narrowing circles, closer and closer to the earth. Then, with the speed of a jet plane, it suddenly swooped onto the gloved fist of a man. From wing-tip to wing-tip it measured more than four feet. The man gently smoothed the feathers of the bird and talked softly to it as he fed it a morsel of meat. Then, with a sharp command and a movement of his wrist, he sent it soaring high in the air.

The bird was a falcon, a close relative of the majestic eagle and the familiar hawk. The man had trained it from a fledgeling to attack other birds, and to come and go at the sound of his voice.

More than 400 years before, in 1558, a party was in progress at Windsor Castle, a few miles from London. It was several months after the coronation of Elizabeth I, and it was the queen's birthday. Gifts from all sections of Britain and from many parts of the world were heaped around her chair.

Suddenly there was a stir in the crowd of guests. A dust-covered courier had arrived from a Spanish ship.

The throng made way for him as he strode directly to where the young queen was sitting. In his gloved hands he carried a large cage covered with a dark cloth. After he had bowed to the red-haired lady in the huge chair, he said: "From his Gracious Majesty, Philip of Spain!"

All eyes stared as he gently removed the cloth from the cage. Then there were gasps of astonishment and delight. For, swinging slowly on a silver perch, was a snow-white falcon, its head covered with a hood of gold. The queen was pleased. For she, as well as most women in the courts of Europe, enjoyed the sport of falconry as much as the kings and their nobles. Only the nobility or people of great wealth could enjoy the sport, for it required immense estates where grouse, pheasants and quails abounded.

Training and using these ferocious birds of prey is said to be man's oldest sport. It was practiced in China earlier than 2000 B.C., and there is a statue in Iran, dating from 1700 B.C. showing a hunter with a falcon perched on his wrist.

Falconry was so popular in the past that even the clergy took part in the sport. Early in the 14th century when some falcons belonging to the Bishop of Ely in Cambridgeshire, England, were stolen, the thieves were excommunicated. In India, when the most-prized falcon of the Maharaja of Bangalore died of old age, the Indian leader was heard to wish it had happened to one of his sons instead. And when Marco Polo, the Venetian traveller, visited China early in the 13th century, he reported that the most-prized possession of Kublai Khan was a special falcon.

Every country has some legends concerning the falcon. The Japanese believed that when one of their mythical ancestors stepped ashore from the boat that brought him from China, he was amazed when a falcon swooped from the sky and perched on his bow. From then on the bird represented success and bravery to the Japanese. Even now it is shown on the Medal of Victory, which their government bestows on its heroes.

More about Falcons

There are several types of falcons, but the one that has been man's sporting and hunting companion for centuries is the peregrine.

One of the ways the peregrine differs from its eagle and hawk cousins is in its "voice box." The sound it makes is a raucous scream. It also has longer and more pointed wings, enabling it to fly at an average speed of 55 miles an hour, and to dive on its prey from great heights at the fantastic speed of 275 miles an hour!

Peregrines mate for life. When courting, they go through astonishing aerial gymnastics, screaming at each other as though doing battle. They are vagrants, building no nests of their own, but taking possession of the nests of other large birds high in trees or on cliffs, forcing less aggressive birds, including hawks (and even eagles) to abandon the sites.

Most peregrines are dark blue-grey, with black stripes across the breast. (The white falcon given to Elizabeth I was probably an albino.)

Peregrines lay three or four eggs at a time, which are hatched by the female, the more ferocious of the couple. She is also larger than her mate, and twice as strong. That's why the female is used in falconry. For her size she is the most cruel and savage creature on earth.

The babies, called "eyasses," require constant attention, taking about six months before becoming strong enough to leave the nest. During that time they are easy prey for two of their deadly enemies, owls and eagles.

Falconry is a sport in almost every country. In the past, and even now in some places, the birds were used to procure food for their owners. However, now the sport usually involves training the birds to strike and hold some inanimate objects that have been thrown or shot into the air (pieces of wood, etc.) to which small portions of meat have been attached.

Training a falcon requires patience, kindness and perseverance. The bird is kept in almost complete darkness, usually in a shed or room where the windows have been covered anc dim electric lights are used sparingly. When taken outdoor the head and eyes are covered with a leather hood. Falcons ar nervous and excitable, and, without the hood, would b difficult to control. However, when the head covering is re moved the bird eagerly attacks anything that moves in the sk

Training the young bird to return and perch on the wrist accomplished by attaching a thin leash to the creature an allowing it to fly for a few yards, then pulling it back to th hand and giving it a morsel of meat. The flights are graduall lengthened, until the bird will return from hundreds of yards.

With patience it is eventually possible to free the bird completely, and have it return to the hand for food. Always, whil feeding, the trainer talks to it. The falcon soon learns to associate the tone of his voice with food, and will return from high overhead when he calls. Some birds are trained to respond to a whistle, or the movement of an arm.

For a long time falcons were scarce in the United States, having become victims of pesticides, especially DDT. This particular spray affected the shells of falcon eggs (as it did the shells of other birds), making them so thin that they broke easily and few "eyasses" hatched.

However, now that DDT has been banned in many areas, falcons are reappearing in increasing numbers. Perhaps in a few years falconry will again become one of the world's popular sports.

Cain and Abel

There is no more awesome sight than watching wide wings spread in the sky, soaring effortlessly, then a dive of tremendous power, ending in grasping talons reaching for and snatching an animal. Then furiously beating wings carry the prey aloft to an aerie in a tree or cliff.

That spectacle of ferocious attack has made many countries —both ancient and modern—adopt the eagle as a representative of national strength.

The very early Greeks believed the bird to be a special messenger of Zeus. However, the earliest record of the eagle as the emblem of a people was in Sumer, an ancient region in the Euphrates Valley. It was then taken as a symbol of strength by Assyria and Babylonia, and then by the Persians and the Etruscans. At last, in 87 B.C., it was officially adopted as the emblem of the Roman Republic.

When Rome fell, the conquering barbarians recognized the bird as the representative of former Roman might, and used it themselves. When Charlemagne became ruler over what had been the Roman Empire, he made the eagle the emblem of the Frankish empire. Later, the Czars of Russia and the German Kaisers kept up the tradition.

Although there are dozens of species of eagles soaring through the skies of the world, when the United States Congress selected one for the new nation's national emblem it chose the "bald" eagle (so called because of the white feathers that decorate its neck and head), for the reason that the bird is found only in North America. The final design of the emblem, passed by the Congress in 1782, shows an eagle with head turned in profile, with a spray of olives in its right claw—denoting peace; while its left talon grasps several Indian arrows—meaning "we are able to release dreadful thunderbolts on any who threaten us."

However, in recent years this magnificent creature of the skies has been disappearing. Many things have been blamed, but there are two principal reasons. One, of course, is the rapid expansion of population, with the result that many once-wild areas have become merely suburban parts of large cities. This has condensed the flying spaces of the eagles, and eliminated many nesting places.

The second and more important reason for the disappearance of the United States' national bird has been the widespread use of pesticides, especially DDT and DDE, a metabolic substance produced by DDT. This insidious poison caused, not only eagles, but many other birds (as we have seen with the peregrine falcon), to produce abnormally thin-shelled eggs. These eggs often broke before hatching, but many were infertile.

This effect on birds' shells, although dangerous to many species, was most devastating on the eggs of eagles. Other types of birds lay four or more eggs at a time, and many have broods several times a year. But an eagle pair will produce only two eggs every 12 months.

More damage has been done to the eagle by pesticides than to other bird species for another reason—most small birds have a diet of seeds and berries, whereas the eagle, being larger and a meat eater, is a victim of the entire ecosystem. That is, as the poison builds up through smaller animals, it becomes a tremendous amount when it finally reaches the larger animals the eagle feeds on, and therefore it is the eagle that receives the greatest and most lethal dose. However, now that many pesticides, and especially DDT, have been eliminated or are being withdrawn from the wilds, eagles and other birds are slowly returning.

It has been estimated by ornithologists that to maintain a constant population eagles must produce .07 young per couple every year. Over the past years the rate has been less than that, hence the decline in the number of birds. However, even when that percentage is eventually reached, it means that a pair of eagles should raise (as percentages are figured) a trifle more than one young bird every two years.

But a pair of eagles almost always produces two eggs each year. What, then, becomes of the second egg? Why doesn't the couple raise two birds? Very often when two eggs are laid, and two eaglets are hatched, only one survives. Why?

The problem has been called the "Cain and Abel" syndrome.

According to an article in the *Smithsonian Magazine,* two young medical doctors, experimenting in the Carpathian Mountains of eastern Czechoslovakia, have discovered the answer to that problem. What they are doing has saved the life of that second eaglet and they have proved the experiment again and again.

Both doctors had always been interested in raptors (large

flesh-eating birds), and especially eagles. And for a long time they have wondered why they have often found one fledgeling, dead, at the base of the aerie. The dead bird never showed signs of being injured or hurt. Why? After studying for a long time the activity in eagles' nests they believed they had the answer, and experimentation proved them to be correct.

Apparently the two eggs in an eagle's aerie hatch at different times, sometimes a week or more apart. Therefore the first born, Cain, has a start in being fed and having the undivided attention of his parents. Abel, the second born, is shoved aside, not by the parents, but by the aggressiveness of his older brother or sister.

The doctors point out that it isn't a case of physical strength, but, instead, is psychological. "The second born *accepts* the intimidation," they say. "Even after a few peckings, although it isn't hurt, the smaller chick simply surrenders. It crawls to the edge of the nest where it can't be brooded or fed. Cain gets the food and attention; Abel either dies of starvation, or exposure, or falls from the nest because of weakness."

The doctors explain that the Cains in eagles' nests aren't strong enough to kill the Abels, but their aggressive attitudes *frighten* the newly hatched into submitting to being ignored. Their solution is to keep a watch on an eagle's nest, and just after the second egg has been hatched (and both parents are away seeking food) to climb to the aerie and remove Abel.

Then they carefully transport the chick in a basket, often by car, and place it in the nest of another type of raptor, usually a kite. The operation is carefully planned. The fledgeling kites, if any, have been removed to the nests of other birds that will accept them. Kites have proved to be excellent foster parents, and seem not to care if the food they bring back to their nests is gobbled up by a strange chick.

The new boarder is allowed to remain in the kite's nest for

approximately two months. Then the Abel chick is large and strong enough to be returned home, where the older brother or sister is no longer able to threaten it. And where both chicks can then be taught by their parents to fly, seek food and survive.

Sometimes the doctors remove the first-born chick and place it in a kite's nest. The experiment seems to work either way. And the interesting thing is that the eagle parents appear to accept the disappearance and reappearance of their offspring without suspicion. The result is that two healthy birds are raised where—too often—only one had survived.

Perhaps if what these two exceptional men have accomplished in a small way, could be multiplied many times, the world's skies would again be alive with eagles.

The Bird with the Radar Ears

There are many kinds of owls, and they can be found in almost every country. The smallest is about five inches in length; the largest more than two feet tall. The snowy owl, one of the giants of the species, can spread its white wings more than five feet.

All owls appear to have stand-up ears that make them resemble alert puppies. But these really aren't ears, they are merely tufts of stiff feathers. The real ears do not have external parts and are concealed by ordinary feathers, yet they are so keen that the birds can hunt and capture small animals at night, merely by sound.

Owls, being night creatures, share the same dismal reputation as other animals that hunt in darkness. Owls can scream, howl, snarl and hiss. But it is their eerie hoot that makes hearers think of evil tidings and tragic forebodings.

But it was not always so.

The Greeks in particular regarded the owl as a token of good luck and a carrier of happy tidings. The bird was the special pet of Athene (Minerva to the Romans), and the statues of that goddess often have an owl at the head. And when, according to legend, during the battle of Salamis, an owl alighted on the mast of the flagship of Themistocles, the admiral, it was assurance that Athene herself was fighting on the side of the harassed Greeks. For a long time afterward the owl was proudly depicted on gold coins.

However, when, later, the Romans identified Athene with their goddess, Minerva, they also transferred responsibility for the owl, a bird they dreaded and detested, to that goddess. Even Pliny, the Roman historian, wrote of owls as being "messengers of death." Pliny also claimed that the deaths of several Roman emperors had been foretold by the screeching of owls—even, according to him, the death of the great Augustus was announced by a brown owl "singing on his doorstep."

Perhaps the sad reputation of owls has something to do with the desolate places in which they choose to live. It might also be due to their disagreeable voices, their peculiar habit of flying so silently, their keenness of sight and hearing, and their strange ability to turn their heads almost 180 degrees without moving their bodies. Perhaps, too, owls have always been associated with witches because they fly at night and can make such frightening noises.

Of course Shakespeare made the most of the dire and horrible reputations of the owl in many of his plays. In *Macbeth,* the good lady, her hands dripping with blood, cries,

"Hark! Peace!

It was the owl that shrieked the fatal bellman . . ."

And when Richard III was irritated by unpleasant news, he interrupted the messengers:

"Out on ye, Owls! Nothing but songs of death!"

The wing of an owl was one of the ingredients in the cauldron wherein the witches prepared their "charm of powerful trouble" in the play *Macbeth.*

It's pleasant to record that legend does give the owl credit for another good deed (besides helping the Greeks at Salamis). It seems that Genghis Khan, during one of his many battles of conquest, had his horse shot out from under him. The great Khan (apparently not so brave as was believed) escaped his enemies by hiding in a bush. As they searched for him an owl alighted on that bush, and the searchers, believing no bird would perch where someone was hiding, went on their way. According to the legend, from that time on, Genghis Khan kept many owls as pets.

The owl, instead of being a dreadful creature to have on the premises, is one of the most beneficial. Farms and gardens would be overrun with rats and other vermin without the owl-guards that keep watch, and whose "radar" ears can detect the slightest movement, and so keep down a pest population.

The Phantom of the Arctic

It sits on a snow-covered hummock in northern Canada, its white feathers blending into the landscape. The feathers ripple in the icy wind that blows almost constantly from the Arctic. The round head slowly turns in a complete circle as bright yellow eyes scan the bleak land for the slightest movement.

When it detects a flicker of brown on the snow, it spreads its wings and takes off almost vertically. It soars only high enough to pinpoint the creature that has attracted its attention. Then, with a swoop of wings it strikes, its coal-black claws digging into the lemming or rabbit. In a few seconds the struggle is over, and the white phantom of the northern skies has another dinner for himself, his mate and his chicks.

It is the snowy owl, one of the most relentless hunters to drop from the sky onto prey. It hunts in the daytime instead of the night, as do other owls. It is an imposing bird, standing two feet tall, with a wingspan of five feet.

The snowy owl doesn't always stay in the far north. There are times when it migrates far to the south. Its movement depends on the supply of its principal food—lemmings. Although it eats small birds, mice and even the large northern rabbits, it depends mainly on lemmings, rodents whose numbers vary from year to year.

In some years, lemmings have a birth explosion, and breed by the millions. However, this often leads to mass suicides and epidemics, when the creatures die in huge numbers.

When lemmings are plentiful, the snowy owls stay north and produce as many as 15 eggs in a season. When lemmings are scarce not only do the owls fly south to find food, but their egg production drops to one or two, and sometimes no nests are made at all. Fortunately, lemming populations run in definite cycles. There are bottoms, as well as peaks, in their numbers, and therefore the number of snowy owls remains almost constant.

The courtship of the snowy owl—usually in April—is one of the most fascinating among flying creatures. According to the Canadian Wildlife Service, each male owl goes a-wooing with a lemming clutched in is beak. After selecting a female, the male, still holding the lemming, performs astonishingly complex aerial gymnastics for her benefit.

After completing his winged acrobatics, he alights near the female and lays the lemming at her feet. He also gives voice to his courting song, a sound that begins in the lower register and ends on a sustained loud howl. If the female shows no interest and turns away, the male picks up the lemming and tries his luck elsewhere. However, if she picks up the rodent, the deal has been made, and they begin to build a nest made of twigs lined with feathers and grass, usually on top of a hummock.

If the lemming-hunting season has been good, the first eggs are laid in early June. The female does all the brooding, while the male busies himself finding food for them both. While sitting on the eggs, the female ignores the snow and wind. Sometimes a blizzard completely covers her and the nest, but she shakes off the snow and remains on the job.

The eggs begin to hatch in a little over a month. If a larger number has been laid (from 10 to 15), there is some time between the appearance of the first chick and the last, and a nest might contain a number of offspring in various stages of growth. The older owlets make room for the newcomers by leaving the nest and crouching around it. But they still are carefully fed by the father, who may have to search some distance for the more adventuresome of the chicks.

In four months the youngsters are ready to join their parents in the search for food in that area, or, if lemmings are scarce, to accompany them south.

When these magnificent phantoms of the north do fly south they often gather at the airports of Canada's largest cities, for reasons unknown. And that often results in some unfortunate events.

Many air accidents have been caused by the large white birds being sucked into engines or flying directly into pilots' observation windows. Since 1959 the damage to Air Canada planes has amounted to more than $3,000,000. Until lately the only remedy was for the security guards at the airports to shoot any snowy owls that appeared. However, not only did the Canadian Humane Society deplore the massacre, but many guards refused to shoot the birds.

Now the airports are using more humane methods. Instead of shooting the owls, they are trapped, then released into the barren lands farther north, with the hope that the lemming population will again have begun an upswing and there will be enough food to keep the mighty hunters satisfied. The new plan has worked, and shots are not heard around the airports.

Now that the Canadian Provinces have laws that fully protect the snowy owl, it is certain that the magnificent phantom of the Arctic will be with us as long as there is snow and ice—and lemmings.

The Invasion of the Starlings

On March 6, 1890, a man named Eugene Scheifflin, committed an act that has made him infamous to airline pilots, and to the sanitation departments of every town in the United States.

For on that day Mr. Scheifflin loosed into the air, in New York City's Central Park, eighty European starlings. Mr. Scheifflin had the bizarre plan of introducing Americans to every species of birds mentioned in Shakespeare's plays. It was Mr. Scheifflin's only attempt to make his dream come true. But it was more than enough.

Since that momentous day in 1890 starlings have proliferated to an alarming degree. They have adapted so eagerly to the American environment that they now are the most numerous bird species in the United States. By 1898, flocks of starlings had become firmly established in the eastern states. By 1946, they had found their way to California. Today they exist in astronomical numbers in every state.

Starlings are aggressive. They take over the nests of other birds—flickers, finches, robins, woodpeckers and even sparrows. They are like visitors who can't be insulted, and soon take over the premises.

Starlings seem to shun the wilds. They are city birds. They love highways, airports, streets, bridges. Wherever people congregate, these unmannerly immigrants make nests, make love, swarm and impede the lives of other birds, as well as people.

Although starlings at times emit an eerie whistle all their own, they are proficient in imitating the calls of other birds, even parrots. Their vocalizations blend into the mechanical noises of modern society. They are the perfect, if unwelcome, city dwellers.

They love the activity around airports to such a degree that at times blizzards of them delay the flights of planes. Investigations have proved that many air accidents have been caused by jet engines and propellers becoming fouled with the bodies of starlings.

These swooping flocks also cause hazards on the huge bridge-arteries that feed New York, San Francisco and other metropolitan areas. They courageously, or stupidly, fly into windshields, and if the bridges support walkways, often it is impossible for pedestrians to use them.

Perhaps because of activity and motor noises, starlings seem to delight in roosting and flying around Army forts. Whatever the attraction, they have become a nuisance that hinders many manoeuvres at military installations.

To reduce the vast starling populations that inhabit these and other places, the Army has tried shotguns, propane cannon boomers (hoping the noise will disperse them), taped starling "distress calls" to scare them, and thinned the trees where the birds make their homes. It has even sprayed the trees with chemicals that will not kill, but are so unpleasant only the most determined birds can stand them. But starlings *are* determined. They manage.

Starlings appear to be impervious to sprays and chemicals that would be deadly poison to other birds. They are a perfect example of a creature, like the cockroach, that can tolerate any change and overcome any obstacle.

Starlings can find footholds on twigs and wires, and the ledges of buildings provide preferred nesting places. Their all-purpose bills can discover food where other birds would starve. They can open trash cans and bags, and will, if in the mood, eat the putty around window panes.

When Charles Darwin announced his theory of the "survival of the fittest" he probably was thinking of the many difficulties nature placed in the path of all species, forcing them to adapt or die. He didn't dream that in time civilization itself would foster the spread of a bird that thrives on man-made barriers. The starling is an excellent example of that new variety of life.

Starlings are like guests who come to dinner, and stay, and stay . . .

Were Ravens White?

In many ways ravens are like most other birds. They are about 26 inches long, and have wing spreads of approximately 36 inches. They produce five to seven eggs each year, and are good parents. They also fly, and they eat.

But in some ways they are *very* different.

They are black. But not the black of the crow. The raven's black has a purplish cast.

They also have the reputation of being birds of ill omen, carriers of bad news. (It does seem strange that, throughout history, they are reported to have been present when unfortunate events happened.) However, that reputation, whether right or wrong, began many years ago, at a time when various gods were believed to rule the earth, air and water. It was also a time when, according to legend, ravens were white.

One of those gods was Apollo, second only to Zeus among the deities of the ancient Greeks. Apollo owned a pet raven that loved him very much. That love was so great that the bird was jealous of anyone who was close to its master. When Apollo fell in love with Coronis, a beautiful girl from Thessaly, the raven, in an attempt to break up the affair, whispered in Apollo's ear that the girl was unfaithful. Apollo was furious, and slew Coronis. Later, Apollo learned the truth, that Coronis was indeed faithful, and his anger was turned against his pet raven. As a punishment he placed a curse on the bird:

"... he blacked the raven o'er
And bid him prate in white plumes no more."

Through the ages other legends have tried to explain why ravens are black. In Italy, the town of Brescia still observes January 30 and 31, and February 1, as "blackbird days," in remembrance of an unusually cold winter many years ago, when the ravens (then white) took refuge in chimneys for warmth, and ever since have worn a sooty plumage. That might be the reason ravens are associated with unpleasant events, and why Edgar Allan Poe called his sad poem, "The Raven."

However there *are* rare legends relating how ravens were of use to people. One of them is still believed.

In the grounds of the Tower of London, one of the oldest prisons and fortresses in Europe, there are small flocks of ravens at all times. But these birds can't fly, for their wings have been cut. The story goes that when the Normans occupied the Tower and were attacked by enemies, they were warned by the loud squawkings of ravens. From that time on, many people have believed that as long as there are ravens at the Tower of London there will always be an England.

Therefore, when any of the Tower ravens die, new ones are obtained immediately and their wings are kept clipped so that for the remainder of their lives they will stay on the grounds. When questioned, the Tower authorities refuse to admit they are superstitious. They explain they are only following tradition. But the ravens remain.

Among employees listed on the roster kept by the Quartermaster of the Tower, the ravens are noted as being "on the strength of the army." In other words, these birds of ill omen are definitely "soldiers of the Queen."

Why Sparrows Hop

Most birds walk or run when on the ground. That is, they move by placing one foot in front of the other.

But not sparrows. They hop.

Why?

Perhaps two ancient Russian legends have a little truth in them. One tells the story that while Christ was hanging on the cross the sparrows chirped "Jif! Jif!" meaning "He is still living! He is still living!" in order to urge his tormenters to more cruelties. But the swallows cried "Umer! Umer!" meaning "He is dead! He is dead!" hoping he would be let alone and suffer no more. For that reason the swallow is blessed, but the sparrow is cursed, and since then has had to hop because its legs are bound with invisible cords.

Another legend relates that when the soldiers were searching for Jesus in the Garden of Gethsemane, the swallows tried to entice the searchers away by their erratic flying. (Swallows still swoop and fly that way today.) But the sparrows flew directly to where the Saviour was standing, and so pointed him out to his captors.

Therefore sparrows are burdened with two curses.

Although there are hundreds of kinds of sparrows throughout the world, the one we are most familiar with—the "house" sparrow—is a relative newcomer to the United States. A pair of these birds was released there in 1852 by an Englishman who wasn't aware that there were native American sparrows. He thought "English" (house) sparrows would make him feel more at home in his new land. That one pair became the ancestors of the millions of "common" sparrows that hop along the streets of American towns and cities today.

All other kinds of sparrows—chipping sparrow, song sparrow, grasshopper sparrow, tree sparrow, hedge sparrow, to name a few—like wide open places. But not the house sparrows. They are city dwellers, and their feathers often show the signs of urban pollution, being covered with soot, grime and grease.

Like the starlings, house sparrows adapt. They build their nests in every available crevice—under air conditioners, beneath roofs, and in the most dangerous and unlikely places, so long as they can be near people and noise. One nest of sparrows was found at the tip of a railway-crossing bar. And there it remained until the young were hatched, even though the long wooden arm was raised and lowered dozens of times each day. Whether or not the family enjoyed the rides is unknown, but the nest did stay intact until the young ones fluttered away.

These remarkable birds even build nests inside other birds' homes, especially the huge ones erected by storks and hawks. But the little intruders try to be as unobtrusive as possible, and make side entrances into the small cubicles they call home, so as not to disturb the larger landlords of the properties.

One of the reasons for the huge numbers of "house" sparrows is the astonishing amount of lovemaking between the males and females. Some birds mate for life, and if the male dies the female remains a widow, or vice versa. But not sparrows. The female never seems to receive enough affection from only one male, so she has many mates.

The result is that she produces four to seven eggs three or four times each season! However, even though she showers her affections among many males, she makes an excellent mother, and lovingly takes care of each brood until the fledgelings leave the nest.

And so, even if sparrows must "hop" instead of walking or running, because of the misdeeds of their ancestors, they do enjoy life—especially the females.

Crows—Our Hidden Friends

It is estimated there are more than 3,000,000,000 crows in the United States, in spite of all the guns, dynamite and chemical sprays that have been used to kill them. But they survive. And by surviving do more good for mankind than they do damage.

For centuries men have done battle with crows. But always the birds have won. Are they smarter than humans? Reverend Henry Ward Beecher thought so. He said: "If men had wings and wore black feathers, few of them would be clever enough to be crows."

The principal reason why crows are so disliked, especially by farmers, is that they are blamed for eating the crops. It's true they do eat some of what's growing in the fields, but the percentage is small, and *without* crows there might not *be* many crops. According to experiments made by some biologists in Pennsylvania, one family of crows will destroy approximately 40,000 harmful insects in one season, among them grubs,

caterpillars and army worms—insects that other birds ignore, and that, if left alone, would make many crops worthless.

On the island of Martha's Vineyard, Massachusetts, for example, one angry farmer offered to pay a bounty for crows killed on his land. Farmers from the area eagerly joined in the massacre and eliminated the birds. However, it wasn't long before the farmer discovered that his pastures were disappearing because white grubs were cutting the grass at the roots. His solution? He posted signs reading NO MORE CROW SHOOTING! But it was almost a year before enough crows returned to control the pests and allow his pastures to flourish again.

This amazing and unappreciated bird belongs to the same family as ravens, daws, rooks, choughs, nutcrackers, jays and magpies. The reason there are so many crows is that they *adapt*. They seem to enjoy the challenge of pollution and other modern obstacles that mean death to their relatives.

Four to six blue-green eggs are laid, and then incubated by the mother for 18 days. The father is the provider, bringing to the nest all types of food, from seeds to discarded parts of sandwiches. Crows are like sanitation squads, cleaning up everything in an area. What they can't eat they use in nest-building.

Crows make excellent parents, keeping their offspring with them until they are almost full grown and have learned the tricks of survival. Crows also appear to mate for life. If a partner dies it's not unusual for the survivor to be accepted as a companion by another couple.

Crows aren't loners. They like to be with others of their kind. There are even crow cities, where hundreds of thousands live in trees as far as possible from man. From these cities they travel in flocks to seek food.

Another thing that helps them survive in a world that dislikes them, is that they will swoop to help another crow in trouble. A wounded crow will utter a cry that will bring a flock winging to his side. They'll remain there until either their friend dies or recovers enough strength to fly home with them.

That distress cry also costs many crows their lives. Hunters tape it, then send its amplified sound for miles, bringing flocks of crows to become easy prey for their guns. In Harrison, Illinois, the result of that call cost the lives of 10,000 crows in one night.

A crow not only "caws," it can make more different sounds than any other bird. It can coo like a dove, cry like a baby, sound like an approaching train, or make a noise like a baby's rattle. It also can be taught to speak. One man in Palm Beach, Florida, had a pet crow that had a vocabulary of 20 words.

Not everyone dislikes crows. The town of Lyndon, Illinois, is known as the "Crow Capital" of the world. It even has a crow festival every year. The reason is that the town had a pet crow named "Rocky." He was everyone's friend, and the town even passed an ordinance forbidding scarecrows. Everybody knew Rocky, and everybody looked after him. But there were exceptions. Death finally came to Rocky at the hands of a farmer who didn't agree with the town's ordinance.

Rocky was given a splendid funeral. Someone made him a steel coffin, and he was laid to rest beneath a $300 headstone, that reads:

Here lies Rocky. The Crow. Everybody's friend.

Never Tease a Robin

Perhaps no bird in all the world is as representative of happiness as is the gentle European robin or robin redbreast. (The American robin, which is really a thrush, resembles the robin redbreast, but is larger.) The robin announces the coming of spring in Europe. And its presence during summer, with its cheery red breast and lilting song, can't help but bring peace and delight to the most saddened heart.

Even its eggs, from five to seven laid by singing mothers three times each year (all pure white with splashes of brown), seem to proclaim to all who are fortunate in seeing them that nothing too wrong can happen in this world.

For almost 2,000 years in many European countries, robins have been among the most popular of birds, not only because of their beauty and song, but because of events they are supposed to have participated in long ago. One story is that a robin, saddened by the cruelties being done to Christ as he hung on the cross, plucked a thorn from his crown, wounding its mouth and beak, and so dripped blood onto its breast. A few lines of a poem describes what might have happened:

"Bearing his cross, while Christ passed by forlorn,
His forehead by the mock crown torn,

A little bird took from that crown one thorn,
To sooth the dear Redeemer's throbbing head,
That bird did what she could; its blood 'tis said,
Down-dripping dyed its tender bosom red."

Still another story explains that the robin's reddish front became dyed by the stain it received in trying to staunch the blood that flowed from Christ's side.

The robin is especially loved in England because of a tale about two children left alone to die in a forest by an uncle who would gain their fortune at their death. A poem of that time goes this way:

"Their little corpses the robin redbreasts found,
And strewd with pious bills the leaves around."

Other poems repeat the sad tale:

"No burial this pretty pair received
Till robin redbreast piously covered them with leaves."

And so, although boys in almost all countries search for birds' eggs, in England the eggs of the robin are left undisturbed. Perhaps it's because of an old proverb in Cornwall that warns:

"They that hurt robins or wrens
Will never prosper, boys or men."

While in Scotland boys are told:

"The robin and the wren:
If ye harry their nests
Ye'll never thrive again."

The moral is, don't tease or hurt a robin, or disturb its nest, if you want to be happy and prosperous. Better yet, don't tease or hurt any bird, and leave their nests in peace. They have enough trouble surviving in a polluted world.

Swallows

Every year at dawn, on March 19, groups of people gather at a small mission at San Juan Capistrano, a village 40 miles south of Los Angeles, California. They come by bus and private cars. They bring picnic baskets, chairs, and even tents, and almost all have binoculars. For they are going to watch for something that has happened at that small mission every St. Joseph's day since 1798.

What they are waiting for, and what they always see, are flocks of swallows that glide out of the sunrise to rest and nest in the turrets of the mission after a journey of more than 7,000 miles from their winter homes in Argentina. The flocks of the small fork-tailed birds number between 5,000 and 6,000, but only 300 to 400 actually make their nests in and around the ancient church.

Why those particular flocks fly to that specific place, and on that special day, no one knows. What is certain is that swallows have arrived at the mission on that spring day since the adobe church was built almost 200 years ago.

Among the more than 80 types of swallows, the most familiar is the barn swallow, and it is this one that has the most prominent forked tail. All swallows are travellers, staying in the north until October, then flying south to Australia, southern Africa and South America. In Argentina, flocks of swallows from many parts of the north seem to arrive at the same time, after flying an average of 600 miles a day. One day (usually at the end of October) there are none; the next day the trees are dark with them and the air is filled with their constant twittering and chattering.

After a long northern winter a swallow is a most welcome creature to see and hear. However, "One swallow does not make a summer." That Greek proverb is to be found in Aristotle's *Nicomachaean Ethics*. Shakespeare also mentions it in *The Winter's Tale*:

"Daffodils that come before the swallow dares,
and take the winds of March with beauty."

Swallows make use of their saliva in nest building, using pellets of mud mixed with grasses and straw, and lined with feathers. The nests are attached to caves, cliffs, eaves, trees and buildings with the saliva-glue.

Usually four to six white to brown eggs are deposited, which the female incubates for approximately two weeks until the eggs hatch. From then on both parents bring small insects to their young. Very often a second brood is raised soon after the first chicks have grown old enough to leave home. Swallows are social birds, liking one another's company. Where one nest is made, dozens and sometimes hundreds will follow, and the constant twittering, chirping and chattering, can be quite noisy.

Swallows are seldom at rest. Like hummingbirds they are constantly in the air, eating insects while in flight, and even drinking as they hover over water.

There are many legends concerning swallows. Mentioned elsewhere in the book is the legend that they tried to protect and console Christ as he hung on the cross. Therefore it is considered bad luck by many to hurt or kill them. Dryden, the English poet, refers to that belief in *Hind and Panther*:

> "Perhaps you failed in your foreseeing skill,
> For swallows are unlucky birds to kill."

One of the strangest folktales is that swallows have somewhere in their interiors two small stones, one red, one black. The red one will instantly cure any invalid; the black will bring good luck as long as it is possessed.

The swallow is concerned with another stone, too. This one the bird is supposed to find on a beach, and it will restore sight to the blind. Longfellow, the American poet, must have thought of it when writing *Evangeline*:

> "Seeking with eager eyes that wondrous stone which the
> swallow brings from the shore of the sea to restore
> the sight of her fledglings."

However, no legend explains how to obtain these marvellous stones without killing the birds.

He Carries the Sky on His Back

The bluebird is like the bringer of good news in a time of woe and trouble. After a winter of gloom, snow and ice, his bright blue plumage and soft melodious voice announce that spring is on its way, that things will be pleasant, and that the world is indeed a good place in which to live.

The bluebird is as American as apple pie. Its lovely blue feathers welcomed the Pilgrim Fathers, who called it the "blue robin," in memory of the birds they knew and loved in their homeland.

This beautiful creature, like the American robin, belongs to the thrush family. A trifle smaller than the robin, it has bright blue feathers on its head, wings and back, and has a red-brown throat and chest. In spring and summer it can be found in all parts of the country. In winter it flies south as far as possible to avoid the cold.

Other small birds—cardinals, orioles and jays—are flashier and flaunt their plumage by drawing attention to themselves with loud noises. But the bluebird is quiet, appearing even to be philosophical, as it sits calmly on a limb. Even its soft voice seems to say, "Dear, dear, think of it. Think of it." It is a bird that represents serenity and peace. It seldom asserts itself.

Not too many years ago bluebirds were among the most numerous of American flying creatures. There were few fields, woods or villages that didn't enjoy the flash of blue as the birds made their nests, especially in orchards. But they have been slowly disappearing. There are many reasons for their scarcity. Pollution and chemicals have killed many of them, of course, but the principal reasons are the urbanization of once-serene areas and the fact that sparrows and starlings have become their enemies. Again, it is an example of the survival of the fittest and most aggressive.

Sparrows boldly take over the few nesting places left in old trees, and starlings make fierce attacks on the nests still occupied, stealing and eating the four to six light-blue eggs, which are left unguarded for only short periods.

Thoreau, who loved all wildlife, made a special note about bluebirds: "Princes and magistrates are often styled serene," he wrote in his journal in 1859, "but what is their turbid serenity to that ethereal serenity which the bluebird embodies?"

Many American Indian tribes considered the bluebird sacred. Because of its azure plumage it was regarded as the representative of the South, and the herald of the rising sun. One ancient Indian saying claims that:

> "Two bluebirds stand at the door of the house in which the gods dwell."

The Indians also have a legend about how the bird became blue. Originally, they say, it was an unlovely grey. But because of its reputation for gentleness it was allowed to bathe in a certain lake of blue water that had no inlet or outlet. It bathed in the lake for five mornings. On the fourth morning it emerged bare, with no feathers at all. However, on the fifth morning it arose from the lake clothed in a coat of bright blue.

Fortunately, bluebirds may not disappear entirely, for those who love them and regard them as truly "birds of happiness" are doing something about saving them. The solution is to make homes for them, homes to take the place of those destroyed by bulldozers and greedy sparrows and starlings.

In many American states concerned people are erecting "trails of boxes," and bluebirds are using them in increasing numbers. But these new homes must be carefully constructed. If the entrance hole is exactly $1\frac{1}{2}$ inches in diameter, and the box at least six inches deep, starlings can't enter or reach down far enough to destroy the eggs or young. If more people care, and more boxes are placed, the bluebird might again bring joy to those who yearn for news that the winter is past and that soon the land will again flourish.

Early in the present century, long before it was feared that bluebirds might become extinct, the U.S. Department of Agriculture issued a booklet in praise of the bird. Part of it read: "This bird has never been accused of depredation upon cultivated crops, or of making itself obnoxious in any way."

Not only is the bluebird a representative of peace and happiness, it does, as Thoreau wrote, "carry the sky on its back."

The Pious Pelican

Everyone has probably heard the verse that says the bill of a pelican "can hold more than its belican." But that's an understatement. The pelican's pouch, hanging from its long bill, can contain from one to two gallons of liquid, much, much more than its "belly can."

The pouch is really a three-layered skin bag. The skin is on the outside. The inner layer is a mucous membrane. Between the two is a thin layer of two sets of muscles running in opposite directions. The pouch can be contracted until it's barely seen, or extended into a huge hanging sack.

It was once believed that the pelican used its pouch to carry *live fish in water* to feed its offspring. That isn't true, either. The bird does scoop up water containing fishes, but then, by contracting the pouch, it squeezes out the water, leaving only the fishes. It swallows these morsels, predigests them, then lets its young feed on the parts of the fish it brings back, or regurgitates, into its bill.

According to some Egyptian legends, pelicans do sometimes make their nests far from the sea and carry water to their young. For this reason, Egyptians call the pelican "the camel of the sea."

One legend about the pelican, a stranger legend than surrounds any bird, is that the female pelican is a representative of salvation through self-sacrifice. Thus, some countries use a picture of the bird as a sign of piety, and rebirth, in their religions. As with all myths, this one began with a very small truth. Male pelicans, although they help guard the nest, and even bring food for the young, sometimes—either accidentally or because of provocation—kill them.

The legend has it that the mother, returning to the nest and seeing her dead children, brings them back to life by pouring her blood over them, and so gives up her own life in reviving them.

It's true that when the mother wings back to the nest with food and drops open her huge pouch for the young to eat, it does appear that her entire breast is red. But it isn't her breast, it is the inside of her immense bill. However, when the babies stick their heads into it and peck at the food it does look as if they are picking at it and releasing her blood.

One ancient rhyme explains:

"Then, said the pelican,
When my birds be slain,
With my blood I them revive.
Scripture doth record
The same did our Lord,
And rose from death to life."

For many years the pelican population along the North American east coast dwindled alarmingly. Oil spills at sea, and chemicals washing from the rivers into the sea, took a steady toll of these magnificent birds. However, now that more attention is being paid to ecology, they are slowly increasing. Once again their nests, cleverly hidden in the reeds of marshes, show the usual two to four whitish eggs. And now many open mouths await the return of their parents with food in their amazing bills.

One of the most thrilling sights is to see flocks of from 10 to 30 white pelicans, flying in a V-formation (which has been copied by airmen). Pelicans beat their wings only to rise into the air. Then they sail to great heights, riding the air currents that permit them to soar for many miles. They break formation only to dive for fish. The flock works in unison, driving the fishes into shallow water where the huge birds can take their time filling their bills with food for their young.

The Sea Wanderer

The albatross is the largest of all sea birds. It is also the most graceful of any bird when in flight. With a wing spread of from 11 to 17 feet it can glide effortlessly for hundreds of miles. Fourteen species of albatross live in various parts of the world; four types have homes in the Pacific Ocean and can be seen along the western coast of the United States, from California to Alaska.

Although full of grace in the air, on land the albatross is one of the most clumsy and ungainly of birds. It comes to earth like a barrel fallen from the sky, tumbling over and over, striking rocks and other obstacles until finally it comes to a stop.

It takes to the air in the same awkward manner. It either runs, or rather wobbles, for hundreds of yards with wings extended until it manages to become airborne; or it takes the easy way, and staggers to the edge of a cliff and drops off into space. Luckily, this perfect gliding machine uses the land rarely, except as a place to court a mate and to breed and raise one youngster each year.

The albatross, like the sea turtle, returns to breed where it was born. It selects a mate by doing an intricate dance, a mixture of the disco and the minuet, and eventually one egg is laid. These birds make no nest, but that one egg is carefully guarded and incubated, with male and female taking turns. If the egg is left unprotected for only a few minutes, other, smaller, birds break the shell and eat the contents. There is no second chance for an offspring that year. The parents do their best to protect that precious egg by building a barrier around it with vegetation and sand.

If the incubation is successful a chick will emerge in two months. It will be guarded and fed for another six months before it is ready to make its first flight.

A young albatross doesn't leap into mating. Until it is three years old the bird appears to have no interest in the opposite sex. From three to five years of age it "observes" its elders as they mate and take care of the young. During that time the young birds will form pairs, and "keep company" until they become five years old, or even older. *Then* they begin a family.

The albatross bears a charmed life among all seafarers, for it is believed that to harm or kill one will result in a life of bad luck. This was a superstition long before Samuel Taylor Coleridge, the English poet, wrote *Rime of the Ancient Mariner.* In this tale the seaman had killed one of the birds, and from then on his life was one tragedy after another. The old sailor thought that telling his story again and again might ease his troubles and remove the curse. But even stopping people on the street and fixing them "with his glittering eye" while recounting his misfortunes, didn't help. He was condemned to trouble forever.

The amazing albatross can stay airborne for weeks at a time. Some biologists claim it even sleeps while gliding. It obtains its food by eating surface fishes and other sea creatures.

However, it is most known for following ships to obtain the scraps thrown over the side. There is on record a report from the British ship, *HMS York,* that an albatross followed it for four days, during which time the cruiser covered 1,700 miles.

The albatross, unlike birds that flap their wings to stay aloft, seldom uses its wide spread of feathers except for small movements to gather speed while soaring. For years aeronautical engineers and flying enthusiasts have studied the albatross, seeking to discover *how* it manages to glide so effortlessly for so many miles. All agree no man can make anything to equal its ability.

The Pirate of the Air

The frigate bird is not only the most powerful flying machine created by nature, it also is one of the last completely independent creatures that roam the skies.

It is a bird of the seas, but it can't swim. It is a fish-eater that obtains its dinners by trickery, flying agility and theft, even resorting to threats of violence to rob other sea creatures of the daily catch which they obtained honestly.

There are five species of frigate birds that soar over the open waters in almost all parts of the world. Some of them have wingspreads of five feet, others measure more than eight feet from wingtip to wingtip.

What makes the frigate bird so powerful a flyer is the manner in which its muscles are distributed. The huge breast sinews that work the long wings make up more than a quarter of the bird's weight. The wing feathers, each almost 20 inches long, make up another quarter. The remainder is skeleton. The tail is long and forked, enabling it to do the aerial acrobatics necessary to chase honest birds and make them give up their prey.

Gulls, cormorants and boobies are so relentlessly pursued that the frightened creatures eventually drop their catches—catches that are immediately picked out of the air before they touch the water, and are then carried to a nest or bit of rock where they are devoured.

The frigate bird is an ancient cruiser of the skies. Early Egyptian and Greek fishermen told tales of how it swooped from aloft to rob their lines of fish. Christopher Columbus entered in his ship's log, September 29, 1492, an account of a frigate bird following his ship when it was two weeks distant from San Salvador.

The male frigate bird is coal black from its bill to its tail. However, it does possess a bright scarlet throat sac that it inflates when protecting its territory or when courting. The female is larger and has a white belly.

Both parents take part in the nest-building, which is almost always on an isolated island, and at an elevation. For a frigate bird, whose feet are smaller than a sparrow's and adequate only for perching, needs a long runway to become airborne, or a place from which it can drop from a height and have space for beating its wings to gain speed.

The female brings to the nest site a variety of building materials, usually twigs. She makes trip after trip and depends on her spouse to do the actual construction. He also, with red throat expanded, protects the nest from marauders. For, although he is the greatest of agile thieves while in the air, he is clumsy on land, and other birds use every wile and trick to steal the material he is guarding. On the rocks and cliffs where frigate birds nest, building materials are scarce, and each twig is precious. At each nesting place there is as much chicanery and double-crossing as there might be were several gangs of bank robbers gathered in one place to share the loot.

Only one white egg is laid, and both parents take turns hatching and caring for it. A fledgeling requires several months before becoming strong enough to seek its own food. It is especially helpless when born, for the huge wings grow quickly, requiring immense amounts of blood, and that makes the chick top-heavy and unable to move on its tiny feet. Therefore, the immense wings of the parents are spread over it, like umbrellas, protecting it from the intense tropic heat.

The frigate bird doesn't always steal its food. Even more than the eagle, it is nature's perfect dive-bomber. From where it glides high in the sky it can detect reflected sunlight from a herring or bass that might swim close to the surface. Then, in a power dive of tremendous speed it falls from a great height, levels off with perfect timing, flips the fish into the air with its long bill and catches it before it falls back into the water. No part of the bird gets wet except the tip of its beak.

Or it silently follows a school of flying fish and, again with perfect timing, swoops and captures one of the creatures as it playfully leaps out of the water. Only the beak is used in these attacks; for, unlike the eagle that grabs its prey in powerful talons, the frigate bird's small feet are almost useless.

If the frigate bird were accidentally to enter the water, it would perish. It not only would be helpless, unable to swim or even lift itself from the water, but it could not use its wings. The oil gland, which other birds use to waterproof their feathers, is almost nonexistent in the frigate bird and therefore its water-logged feathers would force it deep into the water and the bird would drown.

The frigate bird's ability to stay aloft for so long has given rise to tales that it can span an ocean without resting, and that it even sleeps while gliding. These stories have never been proved.

However, the poet Walt Whitman, seeing one of the birds

for the first time, exclaimed, "Thou art all wings!" And in a poem describing the bird, wrote,

"At dusk thou lookst on Senegal, at morn, America."

When the secret is known of how the frigate bird manages to glide great distances, perhaps our transatlantic planes might use precious fuel only for starting and stopping.

Nonetheless, it's unlikely that any future man-made gliders will have one of the birds' most eccentric habits—that is the custom of following and swooping at ships, trying to attack the pennants flying from their masts.

That habit was well known to seafarers. In *Moby Dick*, Herman Melville describes the sinking of the *Pequod*:

> ". . . a sky-hawk (frigate bird) tauntingly followed the main-truck downward from its natural home among the stars, pecking at the flag."

Crazy as a Loon?

In the vicinity of lakes in many lonely parts of the North Temperate Zone there often is heard a noise that makes campers shiver in their sleeping bags. It sounds like the hysterical laughter of a madwoman.

It's the cry of a loon, a strange bird that sleeps by day and roams inland bodies of water by night, in early morning and at dusk. It's a bird that not only flies, but swims for long periods under the surface of ponds and lakes.

The phrase "crazy as a loon" has been with us for a long time, for the erratic conduct and sounds of the bird are often similar to the antics and noises of a demented person. But is the loon really crazy? Or are its ridiculous actions and screaming laughter merely demonstrations of its derision and scorn for the silly humans who try to catch it?

That was the opinion of the philosopher Henry David Thoreau, whose book, *Walden,* has been a classic for more than a hundred years. Here's an excerpt from a journal he kept:

> "The loon comes to bathe in the pond, making the woods ring with his laughter. One early morning as I paddled along the shore a loon sailed over me, laughing loudly. When I pursued him he dived into the water. He came to the surface on the opposite side of the boat and laughed again.

"Once more he dived and I quickly rowed to where ripples appeared on the surface. He arose behind me with a burst of demonic laughter. As I turned he plunged back into the water. He must have visited the deepest part of the pond for he was submerged for a long time. When next he appeared he emitted a long howl mixed with unearthly laughter. This was his *looning*.

"I could only conclude that his insane howls were in contempt of my poor efforts to apprehend him, and that he had supreme confidence in his ability to elude me. When again he submerged I thought how surprised the fishes must be to see a *bird* swim swiftly past them."

There are four species of loons in North America. All are large, averaging 36 inches in length, and they somewhat resemble geese. But they differ amazingly in other ways from their waddling look-alikes. Loons not only fly, and can dive from great heights, they also are, as we have seen, very much at home under the water. Some have been caught in nets more than 200 feet beneath the surface of a lake.

Their nests, constructed of vegetation, are hidden along the shore or even in shallow water. Each female lays two brown eggs, and as soon as they are hatched the offspring make directly for the water.

Although swift in the air and water, loons are awkward on land. Because their webbed feet are placed far back on their bodies (enabling them to dive so expertly), they have to propel themselves on land by pushing forward on their breasts. Perhaps that's why they wail in sadness, and emit hair-raising shrieks, for they might be aware they are among the less attractive birds. Or are they really demented?

But whether because of sadness or madness, their wild maniacal laughter has made many campers fold their tents and steal silently away.

What *Is* the Swan Song?

For many years the world's audiences have been thrilled by seeing a spotlighted stage whereon a lovely ballet dancer, with swan-like headpiece, gracefully, gracefully, slowly, slowly, gyrates to a heap on the floor, depicting the "dying of a swan."

Luckily, the dancer isn't required to sing a "swan song," for, according to ornithologists, there isn't one. The bird does make a sound, but it's a loud whoop or a trumpet blast that certainly doesn't indicate approaching death.

However, the myth lingers that the beautiful creature does sing a melodious song before expiring. The expression "swan song" has been in our language for centuries to describe the end of something. We even say someone has given his "swan song" when he commits a crime, fails in business, or makes a last appearance.

Perhaps the legend began with Aristotle, who wrote:

"Swans have the power of song, especially when near the end of their lives, and some persons sailing close to the shores of Libya have met many of them in the sea singing a mournful song, and afterwards have seem them die."

Others, who should have known better, have kept the fancy alive. For example, the poet Byron moans:

"Place me on Sunium's marbled steep,
Where nothing save the waves and I
May hear our mutual murmurs sweep;
There, swan-like, let me sing and die."

Shakespeare mentioned everything "in heaven and earth" in his many plays, and, of course, he didn't neglect the "swan song." These two lines are from *Othello:*

"I will play the swan and die in music.
A swan-like end, fading in music."

In olden times a young swan, or cygnet, made a meal "fit for a king," and in England no subject could possess a swan without a license from the Crown. In modern times, swans, young or old, are no longer eaten. Their place on the dinner table has been left to the goose, a relative.

In earlier times black swans were considered to be an impossibility. So much so that the Roman poet Juvenal, who didn't think very much of women, described a virtuous one as "A rare bird on earth, and very much like a black swan."

Then, in 1697, the Dutch navigator, Willem de Vleming, visited the southwestern coast of Australia and captured four black swans. The bird is now the armorial symbol of Western Australia. When the birds were taken back to Europe, they bred and produced offspring rapidly. Now there is hardly a lake, pond or park in the world that doesn't possess at least one of these beautiful birds.

But whether it sings or croaks, is black or white, a swan moving apparently without effort in a calm and stately manner over the water, is one of nature's most stirring sights. As long as swans are protected (as they are in every civilized country) and as long as they continue to produce five to nine pretty olive-grey eggs every year, these exquisitely graceful creatures will remain on earth.

Nature's Biggest Mistake?

If, as has been said, a camel is a horse put together by a committee, then surely an ostrich must have been assembled by an entire congress, with every member disagreeing violently.

To call an ostrich ugly is to compliment the bird—for there is no word that adequately describes this collection of unsightly, awkward and unpleasant bits and pieces of skin and bone. However, it is a bird. But it can't fly. It can run, though, galloping over the ground at a speed in excess of 35 miles an hour; the fastest land-bound creature with the exception of the cheetah.

The ostrich is the world's largest bird, standing about nine feet tall and weighing an average of 350 pounds. The thin long legs appear to be draped in saggy, baggy pink pantyhose. The barrel-shaped body is covered with drab brown feathers. The plumes of the tail, often hidden by the brown feathers, are the most pleasing things about the bird. These graceful white feathers nearly caused the ostrich to be eliminated from the earth. In the 1800's they were used as hat and dress ornaments by women in all the "civilized" countries.

The huge demand for the plumage caused wanton killing of wild ostriches. The result of these massacres was to make the feathers worth their weight in gold.

Then, fortunately, it was discovered that ostriches could easily be raised in captivity, and ostrich farms became common in parts of Africa. The feathers can be gently pulled from the bird with no harm to the creature, which immediately begins to grow new ones.

The long neck of the ostrich is probably its most repulsive feature. It resembles a long reptile, with a small head perched on top. The belief that the ostrich buries its head in the sand to avoid being seen isn't true. However, the bird does crouch and extend its neck along the ground when it wants to hide.

The long neck of the ostrich is an advantage, for it is like a periscope with two keen eyes at the top, allowing the bird to see great distances. Other animals depend on that periscope-neck, for the ostrich can warn them of enemies long before they come close enough to be dangerous. The ostrich itself need fear no creature, except man. Its long legs and hard hoof-like claws can deliver blows strong enough to kill a lion.

At mating time the ostriches' deep booming voices can be heard for miles. The courtship is just as strange and awkward as the bird. Even the creature's egg-laying is odd. The male digs a whole in which his wives (he often has several) lay eggs. Usually 15 eggs will fill the cavity, but the females continue laying more and more, often making a pile of 50; so many that eggs must be kicked off until there is room for the male and females to take turns sitting on the remaining ones.

The eggs are huge, seven or eight inches long, and weigh three or four pounds each. The *Guinness Book of World Records* claims an ostrich egg can support a man weighing 252 pounds without breaking. The book also states it requires one hour to soft-boil an egg and four hours to make it hard-boiled.

Ostrich eggs are not only a popular food in Africa, but the empty shells are used to carry water. When plugs are placed in the holes made by emerging chicks, or when the eggs are opened for cooking, the shells become excellent canteens, and numbers of them are often carried in net bags on long journeys.

Many ostrich shells, painted with scenes and portraits, have been unearthed in Etruscan and Greek tombs. In Constantinople ostrich eggs were suspended in the mosques, and many are still hung in Spanish churches close to altars. The legend is that the eggs of the ungainly bird represent patience and faith. Perhaps the creature shows patience waiting for its looks to improve, and faith that the miracle will eventually happen.

It is true that the ostrich, which resembles a collection of assorted bird parts on stilts—because of the food and feathers it produces, and its rôle as lookout for other animals grazing near it—does contribute to the health, beauty and safety of others. And maybe, just maybe, it might not be nature's biggest mistake, but a blessing in disguise.

The Impossible Bird

Kiwi seems a strange, even impossible word. But it is no more strange or more impossible than the bird it describes.

The kiwi is one of the ugliest of birds. It can't fly—its naked, small and useless wings are hidden under a mat of thick hair-like feathers. Its head, with a long beak, at the end of which are nostrils, resembles that of an anteater. Its skin is so tough that shoes have been made of it.

Yet this unattractive and clumsy bird, standing a foot tall and weighing about seven pounds, is not only the national emblem of New Zealand, but its ugliness is advertised on that country's stamps and currency. Even the soldiers from that strange island who fought valiantly in World War II proudly called themselves "Kiwis."

Why? In a land where beautiful, but relatively "normal," birds abound, why has this unsightly unbirdlike creature been selected to represent an entire people?

The original inhabitants of New Zealand, the Maoris, gave this weird bird its name because of the shrill whistle it emits that sounds like *keeee-weeeee*. And these ancient islanders made use of the outlandish bird in one important way. The long cloaks worn by their leaders were made almost entirely of the hair-like feathers and skins of kiwis. Some of these cloaks, hanging in New Zealand's museums, are more than 200 years old, yet look as if they were newly made.

It was the coming of white settlers to New Zealand that, at one time, made the kiwi almost as extinct as the dinosaur. These English pioneers, homesick and lonely, in a land where there were strange animals, flowers and birds, imported a few rabbits so they could see and hunt *something* familiar.

However, the rabbits multiplied. They became pests. They soon were everywhere, eating the crops and were even underfoot in the towns. So, ferrets and stoats were imported to control them. That was a second mistake. These new inhabitants must have thought they'd found a ferret-stoat utopia. Rabbits were hard to catch. But, lumbering along the ground in this strange land were creatures that could neither fly, run, nor see very well. (Ornithologists claim kiwis can see barely two feet in front of them.) The result was a kiwi massacre that made the ungainly bird almost as rare as the dodo.

Of course laws were quickly made, beginning in 1908 and reinforced with new rules and penalties every year. Now it is forbidden for anyone to possess stuffed kiwis, their feathers or their eggs, or any items made from them. It is also against the law to export kiwis.

There are so many strange facts about these odd birds that some people might find it hard to believe they exist. They sleep during the day and become active at night. Instead of making nests they burrow into the ground like rabbits. They eat only worms, and each adult consumes about a pound and a half every day. Their sense of smell is so acute (making up for their poor eyesight) that, as they amble through the forests at night with their beaks almost touching the ground, they can smell a worm several inches below the surface.

The egg laid by a kiwi is as curious as the bird itself. It is huge, almost the size of one laid by an ostrich, and often weighs one quarter of the weight of the bird. The egg is so large in relation to the size of the female that, as the time nears to give birth, she staggers around as though drunk, in an effort to keep her body from dragging along the ground.

The male kiwi is the most henpecked spouse of any species. After the female has laid her egg, the male takes on the duty of incubation, sometimes sitting on the egg for from 80 to 100 days. During that time he loses so much weight (for often the female forgets to bring him worms) that he is only a trifle larger than the chick that finally hatches.

And very often after he has brought the egg to life, and has had several weeks of rest, she produces another, and his ordeal begins again. One female laid five eggs during a six-month period, and the male sat on them for a total of 200 days. According to records kept of that particular couple, it was more than six months before the male regained his health.

Although kiwis are wards of the state, a few private hatcheries are allowed to raise some for study. One of these is the Hawkes Bay Acclimatization Society, near Napier, New Zealand. The curator there, F. D. Robson, has had some astonishing experiences with the unbirdlike bird.

At one time Robson had to amputate a kiwi's leg in order to save its life. Knowing the creature would have to be mobile to be able to search for its daily or nightly food, the curator constructed a pegleg for the bird out of a piece of hollow bamboo and attached it to the stump with rubber tubing. It was successful, and for years "Pegleg Pete" covered the ground

as ably as the rest of his companions, and lived a full life, even hatching the eggs his female companion blessed him with.

The reaction of most people who are confronted with a kiwi for the first time is similar to that of a farmer in England who saw a giraffe at a fair. After studying the creature from various angles, he turned away, saying, "There just ain't no such animal."

But the kiwi, impossible as it appears, is alive and doing well in New Zealand.

The Emperor Penguin

Out of the swirling mist that shrouds an ice field in Antarctica, a distinctive figure jumps from the frigid water onto the slippery surface. Very often this figure immediately falls flat on its face, then skids on its stomach for several yards before climbing sedately to its feet. Out of the cold greyness a second creature appears, then a third. Soon there will be dozens. Later there will be thousands.

They are emperor penguins, the largest of that strange family of birds that, instead of flying in the air, prefer to swim and live in the coldest waters of the earth.

The emperor penguins average four feet in height, and all are dressed alike in white shirts that seem to trail on the ice. Their backs and wings are covered by what appear to be natty black coats. If they wore homburgs and carried rolled umbrellas and attaché cases, they might be small diplomats on their way to important appointments.

These strangely feathered birds do have an appointment, the same one they have had every year at that time and place. But why at that season of the year, and on ice fields that surround the most desolate and coldest part of our planet?

The axis of the globe we live on is inclined towards its orbit; therefore, when the Northern Hemisphere is turned toward the sun for six months, giving us our spring and summer, the Southern Hemisphere is experiencing autumn and winter. However, the names of the months remain the same, and so when, in July and August, it is hot in the Northern Hemisphere, it is the dead of winter in the Southern Hemisphere. And this is the time when the emperors mate.

The habits of the emperor penguins remained a mystery for so long because scientists could not believe that their courting, mating and birth season was exactly opposite to that of the majority of creatures, who create new life in spring and summer.

Penguins have been known to us for more than 400 years. When Magellan sailed around the tip of South America (Patagonia) in 1520, he made a note in his journal about the "strange geese" he had seen.

In 1594, Sir Richard Hawkins, on *his* voyage around the world, found his ships short of provisions as they rounded Cape Horn. According to his journal the crew made good use of the "geese" Magellan had noted 70 years earlier:

> "Our beefe having been eaten, we could not have survived without the strange fowles, which we consumed fresh, and also salted."

Those were Patagonian penguins. They still inhabit the islands at the southern tip of South America, and although they were almost eliminated by wanton killing, strict laws now protect them, and once again they number in the millions.

Patagonian penguins differ from emperors in several ways. They are smaller, and their coats are a very dark brown instead of shiny black. Also they make their nests by burrowing into the ground, while the emperor makes no nest at all.

However, it was the Patagonian penguins that led to the discovery of the emperor. When Captain James Cook made a voyage to Patagonia in 1775, he had with him an artist named Forster. Among the many birds this artist sketched were numerous penguins. Much later, in 1884, when the drawings were finally carefully studied, one of the penguins appeared to be different from the others. It was thought to be a freak, and no further thought was given to the matter.

It wasn't until 1902–03 that the emperor penguin was seen alive. Dr. E. A. Wilson, who accompanied Robert Scott on an Antarctic voyage, not only saw emperor penguins, but discovered several of their rookeries. However, he didn't have much time to study them, and what he did see he refused to believe. Why would creatures deliberately court, mate and give birth in the dead of an Antarctic winter? He decided he had been mistaken. He *had* seen the handsome penguins, but what he had assumed were chicks must have been something else.

When Dr. Wilson returned to the Antarctic with Scott in 1911, in the dead of winter, he took the time to study the emperors in their rookeries, and became the first scientist to prove that these magnificent birds' breeding habits were completely opposite to those of all other creatures.

When sheets of ice form on much of the Antarctic Ocean in March—the beginning of the Southern Hemisphere's long winter—the emperor penguins leave the water, ready to mate and raise a family. The time for eating is past. Both males and females have layers of fat beneath their white and black feathers. Nor will the males eat again for at least four months.

Males and females waste little time, getting down to business at once. Males stroll among the females, looking them over.

When one strikes the fancy of a male, he stops directly in front of her and begins to sing. The song isn't the most melodious in the bird world. It begins in the lower registers and rises to a loud, vibrant tone. After delivering his song he waits for a few minutes. If the female makes no response, he moves along until he finds another to his liking and again makes his strange courting noises. When a female does respond with some welcoming sounds, the pair moves off into the wintry mist.

Early in May, as the cold becomes more intense, the egg arrives. There is only one, and the female deposits it directly on the ice. Both parents examine it carefully. Then, satisfied it is as ordered, the male gently lifts it from the ice with his feet and tucks it under him to incubate.

When the female is sure all is well, and that the egg is safely tucked away under her mate, she makes her way to the open sea. The male remains behind, incubating. For two months she will swim in the cold water and eat her fill of small sea creatures. She has lost some weight producing her egg, but she quickly regains her fat, and from then on she eats in order to build a reserve of food for the future.

Back at the rookery the males have stoically stood without moving, the precious eggs carefully guarded and heated by their feathers and flesh. For the two months of incubation they have been without food. They have passively withstood blizzards and frigid blasts of wind. They have become thinner, and their feathers have a dull and tarnished appearance.

But by the end of June or early in July, the coldest months of the Antarctic winter, their relief arrives.

The females return, their stomachs distended with food, their feathers shining with health. As soon as they reach the rookery they begin looking for their mates. This isn't easy, for all the males look alike. However, the female stands in front of one after another and sings. Again and again she passes among them, until finally one of them answers her song. It is her mate, who hasn't had food for almost four months.

From her distended stomach, the female regurgitates some food for her mate, and then takes charge of the egg, tucking it, as he did, into the folds of her abdomen. The hatching takes 48 hours, and the chick, about six inches tall, immediately begins making chirping noises. The mother feeds the offspring by bringing up from her stomach bits of food.

The male waits only long enough to be sure the female and the chick are well before taking his turn going into the sea. He will remain away for two months, feeding to regain his strength. When he returns he goes through the same procedure of finding his mate and chick as did the female after her vacation.

From then until spring the male and female take turns going to the sea for food. When one is away the other stands guard, for many emperor females have lost their young, through accident or cold, and they try to steal any chick they can find.

By this time it is October, and spring is approaching. The ice sheets are breaking up, and the cold is less intense. It has become time for the emperors and their offspring to enter the sea.

Where they go during the Southern Hemisphere's summer no one has yet discovered. All that is definitely known is that in March, when ice again forms over the sea, and cold mists settle over the barren scene, one after another the emperors will climb out of the frigid water and begin the courting and mating rituals that assure that these splendid birds will continue to amuse and astonish all who are fortunate enough to see them.

How the Hummingbird Lost Its Voice

When the Pilgrims stepped ashore in Massachusetts in 1620 the Indian chiefs who greeted them wore strange, small, glittering ornaments in their ears. They were stuffed hummingbirds.

Cortez, the Spanish conquistador, was astonished when he met Montezuma in Mexico, to see that the Aztec emperor and the nobles of the court were arrayed in long cloaks that shone brilliantly when the sun struck them. When a similar cloak was presented to the Spaniard he was amazed to find it made entirely of hundreds of hummingbird skins.

These smallest of all birds (the smallest weigh only 1/10th of an ounce) have several times been almost obliterated from the earth. But each time they have survived. In the late 19th century, when it was the fashion to decorate ladies' hats with feathers, hummingbirds were preferred, and millions of them were killed. In one year, 1860, more than 500,000 hummingbird skins were auctioned in London. But styles change, and when feathers fell out of style, the numbers of hummingbirds slowly began to increase. Also, many enlightened countries, especially the United States and England, passed laws making it illegal to trade in the skins of hummingbirds.

But the small bird has had to overcome other obstacles. It lays only two eggs each year, the size of pearls, and these are the prey of larger birds and squirrels. Also the destructon of forests where they made their nests and the use of pesticides have taken a toll.

Until Columbus and other Spaniards took some of the lovely creatures home, no European had ever seen one. For hummingbirds are known only in the Western Hemisphere. More than 300 varieties have been identified, mostly in Central and South America. Those that do make their homes in North America migrate to warmer temperatures in the winter.

The hummingbird is one of the swiftest of all birds, and flies in the most intricate patterns. Its average speed is 55 miles per hour. It hovers. It flies backward. It dashes, then turns in a flash and darts into bushes, plants, trees and flowers, leaving behind only a memory of sudden brilliance. A hummingbird's wings beat at the astonishing rate of 75 to 90 times *each second.* The bird's speed and dexterity are due to the strong muscles that are attached to the wings, and which account for most of the body weight. Curiously, the legs of the hummingbird are so short that the bird cannot walk or climb as other birds do.

It flashes so much brilliance when it flies because the tip of each wing is made up of refractors or reflectors that catch light exactly as do the facets of a diamond. The fantastic speed of the bird's wings make these reflectors, when hit by the sun's rays, seem to convert the tiny bird into a darting gem.

The hummingbird rarely makes any sound other than the faint "hum" of its beating wings. The reason, so a legend explains, is greed. The bird has an extensible tongue, permitting it to reach deep into a flower to obtain the nectar. Many thousands of years ago, according to the legend, the hummingbird was so voracious it swallowed too much nectar and lost its voice in the flower. And that's why the bird dashes from bloom to bloom so swiftly—it is searching for its lost voice. Actually, the hummingbird emits a feeble chirp on occasion, but few human beings have ever heard it.

Another ancient legend claims that the hummingbird hibernates, instead of flying south to avoid the cold. This, of course, has been proved incorrect by ornithologists, but an old book of folklore explains that the hummingbirds

> "live by the dew and the juyce of flowers. They die or sleepe every yeere in the month of October, sitting upon a little bough in a warme and close place. They wake againe in the month of April after the flowers be sprung, and therefore they call them the revived birds."

Perhaps one day the hummingbird will find its voice in a flower, and the sound will be lovely to hear. Until then it is one of the smallest of the world's natural wonders.

The Bird That Sings with Its Wings

Sometimes in April and May farmers stop doing their chores to stand and listen. Country children tramping through the fields often stare at the sky. And marsh hunters, knee-deep in water, turn their heads to listen to one of the sweetest sounds that ever drifted from the clouds.

What they hear is the courting song of the common snipe, formerly called Wilson's snipe, named for Alexander Wilson (1766–1813), the Scottish-American ornithologist who made it a life study.

Snipes belong to the sandpiper family, which includes stints, curlews, godwits, woodcocks, dowitchers and turnstones. However, snipes are different from their relatives because of the sounds they make.

The snipe is a marsh bird, less than a foot long, brown in hue, with a black-striped back. It is a pleasure to watch it rise swiftly from the edge of a pond or marsh.

The snipe is a game bird, and at one time was the most-sought target along the eastern seaboard of the United States. One hunter kept a record, and claimed he had killed 69,000 of them during a 20-year period. A. C. Bent, in his *Life History of American Shore Birds,* claims the snipe was hunted more viciously than any other game bird. There were two reasons—its flesh, richer than that of the quail, is delicious, and its erratic pattern of flight make it a challenge for hunters.

Fortunately, before the bird became totally extinct, laws were passed prohibiting it to be hunted anywhere in the United States. And so, slowly, snipes again are becoming a pleasure to see, and especially to hear.

For a number of years it was a mystery how this bird made the lovely sound that fell from the sky to startle and please those fortunate enough to hear it. For, when on the ground, or rising in an astonishing burst of speed when startled, the snipe emits only a harsh "scaip! scaip!" Then some ornithologists on a field trip became certain that the music descending from above was from a bird flying high over their heads. They began to investigate and experiment with ways the bird *might* be able to produce that startlingly beautiful chord.

For it *is* a chord. A two-toned harmony. A sound different from the customary chirps, whistles or melodies of other birds. What the snipe makes is *music.* But how?

One scientist attached the tail feathers of a snipe to a cork, the cork to the end of a fishline, then whirled it through the air. That did produce part of the sound, or chord. The second tone was discovered much later by other scientists who experimented with the vibration of the snipe's wings. The two sounds together made up the lovely chord that seemed to float from the sky.

Apparently at courting time in the spring, the male snipe soars to great heights. Then, circling until he spies a hen on the bank of a pond or in a tree, he begins a series of spectacular dives. It is during those "power" dives that his tail feathers and vibrating wings combine to make his "courting" song, and if others besides the female hear it, they are as fortunate as she.

The male continues his fantastic dives and recoveries for hours. The female never rises to meet him. When he is satisfied that he has charmed her with his symphony, he makes a final dive and joins her in the silent and secluded marshland.

Later, in a grass-lined depression in a tufted dune, or tussock, four pale olive-brown eggs spotted with black will be deposited. In a few months the young will be strong enough to join their parents when they migrate south for the winter.

But not until the next spring will the young males send their heavenly "courting" song drifting to waiting ears on earth. That sound is lovely enough to wait for—and to appreciate.

A Most Noble Bird

"A most noble bird!" So cried King Henry VIII as, no doubt, he tossed a drumstick over his shoulder to his waiting staghounds.

The king was dining on turkey, a bird new to England. It had been brought from Spain, and was a descendant of the birds taken home by the first Spanish explorers in the New World. The king so loved the meat that for a number of years all the turkeys bred in Britain were reserved for the royal table. However, by the end of the 16th century so many of the birds were being raised that the ban was lifted.

Domestic turkeys were among the livestock the English settlers brought to America. How surprised they must have been to discover that wild turkeys were common in New England. Domestic turkeys originated from wild birds in Mexico, and had been partially domesticated by the Aztecs long before the Spanish voyages of conquest.

Wild turkeys are still a "game" bird in many states, from Wyoming, across the south and up the eastern coast to New England. At one time they were hunted almost to extinction, but with habitat management, and controlled hunting seasons they once more are becoming numerous.

The name "turkey" is misleading. Hundreds of years ago the bird was confused with some other fowl that had been introduced into Europe by travellers coming from the country of Turkey.

Modern turkeys, raised in flocks for food, would hardly recognize their wild ancestors. Years of careful breeding have created a larger, broad-breasted bird that, prepared properly, is a gourmet's delight.

Wild turkeys roost in trees at night, but they feed on the ground, consuming mostly insects. The male bird struts and prances and doesn't seem to be satisfied until he has a harem of five or six females. Each hen lays up to 18 eggs, which hatch in about four weeks. With just a minimum of protection the bird will continue being abundant in state- and nationally-owned wild areas of America.

This handsome bird, a true native of the Americas, caused a great amount of argument when, in 1782, the Congress of the United States met to decide what device to use on its Great Seal. The eagle was finally selected. One reason, as we have seen in an earlier chapter, was its association with the Roman and other empires. However, there were many members of that Congress who voted not to use the eagle. They felt it might represent too warlike an attitude for a nation whose founders had travelled thousands of miles to seek peace and freedom from oppression.

The second choice of a bird to represent the new country was the turkey. Benjamin Franklin spoke on behalf of this bird, because it really was a *national* bird, being found in the wild only in the Americas (while eagles are found in most parts of the world). He also listed its good qualities, stressing that it had as much courage as the eagle, "and would attack any redcoat that entered its barnyard!"

Franklin also spoke strongly against choosing the eagle, saying:

> "He is a bird of bad moral character; he does not get his living honestly; you may have seen him perched on some dead tree, where, too lazy to fish for himself, he watches the labor of the fishing-hawk, and when that diligent bird has at length taken a fish and is bearing it to its nest the eagle pursues him and takes it from him. Besides he is a rank coward; the little kingbird attacks him boldly. He is therefore by no means a proper emblem."

Franklin lost the argument. The eagle became, and has remained, the national representative of the United States.

However, no one who has sat at a table on Thanksgiving or Christmas could help seconding King Henry's description of the turkey. It is, without doubt, "a most noble bird!"

The Vanity of the Peacock

"An angel's feathers, a devil's voice, and the walk of a thief."

So is the peacock described in Indian and Chinese folklore. In other words, beauty often masks things that aren't pleasant. An ancient poem mentions two faults of the strutting bird:

"The proud sun-loving peacock with his feathers,
Walks all alone, thinking himself a king.
And with his voice prognosticates all weathers,
Although God knows how badly doth he sing;
But when he looks downe to his base black feete,
He droops and is ashamed of things unmeete."

It's true, although the peacock does strut like a king, head-plumes held high, with his beautiful fan-tail spread, he really is like a woman in a glittering gown wearing army boots.

It is also true that the voice of this spectacular bird is a harsh scream. In olden times it was noticed that the peacock's mating call coincided with the rainy season. Its raucous orders to peahens always seemed to announce rain. When that terrible voice was silent there was good weather. Hence the peacock became a weather prophet.

In its original home (India, China and Malaysia) the male peafowl's brilliant plumage, especially the startlingly attractive "eyes" on the ends of its tail feathers, led to its association in mythology with the sun and the rainbow. In Asian jungles, where it roamed freely, its discordant scream heralded the approach of tigers, leopards or reptiles. For this reason, and also because it apparently foretold the coming of rain by dancing and screaming (when all it was really doing was trying to attract a female), it became a bird of magic, and was protected by law.

In China, during the time of the T'ang Dynasty (8th century A.D.) not only was it a crime to harm a peacock, but old chronicles record that

> "thousands of districts paid tribute in peacocks, because their feathers were required by the state, not only as decorations for the imperial processions, but for designations of official rank; for peacock feathers were bestowed upon officials, both military and civil, as a reward for faithful service."

As befits the mate of an Oriental monarch, the peahen walks demurely in her master's shadow. She is drab by comparison, and doesn't seem to care (not possessing her master's vanity) about her ungainly feet. But it is the peahen who builds the nest, lays the six to 12 eggs, incubates them by herself, and takes care of the chicks until they are old enough to leave home.

Even today the peacock is regarded as the most patrician and stately of birds, and many an estate and zoological garden has these strutting, preening, many-hued creatures.

According to owners of peacocks, the horrible scream of the bird still announces rain. Whether it still foretells the approach of dangerous animals is unknown, for not many tigers and leopards roam private gardens.

Birds as Pets

Lonely? A shut-in? Or an outgoing lover of life?

If you belong to any of these groups pet birds can increase your enjoyment of living, in many ways.

Where and when these lovely creatures were first taken into caves and homes as living decorations and for the pleasure of their songs, is unknown. However, the most ancient cave and wall paintings, and our oldest records, describe birds as being important in the lives of people.

They still are.

Birds are enjoyed as pets throughout the world. Kings and queens, and others of great wealth, have had, and still have, immense aviaries where birds swoop and sing, mate and raise their young for the enjoyment of only a few fortunate observers. However, most large cities today have in their zoos various enclosures where birds are on display for the pleasure of many.

But everyone, rich or poor, can enjoy the companionship of one, or a few, bright-hued birds, no matter how small the room, apartment or house. Birds make ideal pets. They occupy little space, require relatively little attention, and need not be taken for walks. In return they will fill a room with bright plumage and song.

With the exception of macaws, cockatoos and other large parrots, pet birds are small. Canaries, budgerigars ("budgies") and various types of small parrots and parakeets are the most common.

A species that is particularly popular, and the easiest to train and care for, is the canary. This beautiful bird is a native of the Canary Islands (hence its name) and the island of Madeira, where it forms flocks that almost cover the trees at certain times of the year.

The canary was first domesticated in Italy in the 16th century, and since then has become familiar in pet shops throughout the world. Its basic coloration is bright yellow. However, careful breeding can produce birds of almost any tint, and even ones with rainbow hues. Captive canaries are able to raise many offspring, producing three and four broods (of up to six eggs each) every year. Wild ones raise even more. Only male canaries sing (the best voices are found in the "Rollers" from Germany); females emit chirps.

The next in popularity is the budgerigar. This small (about the size of a canary) parrot-like bird is a native of Australia, where many can still be seen in the wide expanses of the "Outback." Budgies are often referred to as "love birds" because of the affection they show toward one another. But they shouldn't be confused with the true lovebirds (among the many small birds sold in pet shops) that come from Africa and Madagascar. Budgies are basically green, but by selective breeding they now come in a wide range of coloration. Budgies can also say a few words.

For the amateur bird-lover there is a great variety of parakeets and parrots, which form the third most popular group of

cage birds. (The name "parakeet" is applied to numerous species of small parrots.) All parrots have a small hook at the end of the beak and have feet with two toes pointing forward and two backward. And many can be taught to repeat words. However, care should be taken about *what* words are taught, for sometimes—parrots and parakeets having no manners—what they say can be embarrassing. Coming from South and Central America, New Guinea, Australia, New Zealand, Africa, India, and various other tropical areas, these birds have an amazing range of bright and beautiful plumage.

Among the largest and showiest parrots are the macaws, natives of tropical America. Often kept as house pets, these birds are best suited to a large aviary. Otherwise, they should be kept on stands since they are likely to damage their long tails on the wire of cages that are too confined. Macaws have enormously powerful beaks and are quite capable of chewing the cross-bar of their stands to pieces. The same beak can deliver a serious wound, so you should be careful to obtain a bird that has been handled gently and affectionately.

Homes for pet birds range from standard wire and plastic cages, to enclosed flying areas that are built in garages or attics. However, the most common is the cage that can be hung on a wall bracket or heavy stand. The cage should be large enough to permit the bird to stretch its wings fully without striking the sides. The necessary additional equipment for housekeeping—pots for water, food and grit—should be easily available from outside the cage.

The cage itself should be easy to move, not only for cleaning, but to keep it away from bright sunlight and safe from other household pets—cats and dogs. It also should be kept clear of draughts and heating units. Birds like an even temperature; too much heat or cold will result in loss of feathers and illnesses.

When a small bird, a canary or budgerigar, has become accustomed to its cage, and especially to its owner, it can be allowed freedom to fly around the room. Of course precautions should be taken—doors and windows closed, and open fires and other dangers adequately shielded by guards.

The bird should not be forcibly taken from its cage, but, with the small door open, permitted to take its time before venturing into the unknown territory surrounding its home. The best time for a flight is immediately before feeding. The bird will then return quickly for its food. In time, and with patience, the bird will eagerly take its morning or evening flight.

Larger birds, such as parrots and parakeets, after becoming accustomed to a cage, can be transferred to "T"-shaped stands. At first the bird should be tethered to the cross-bar by a thin chain attached to its leg. However, after it has become thoroughly accustomed to the arrangement, the tether can be removed for long periods or permanently. The food, water and grit containers should be attached to the bar, and around the base of the stand should be a metal or plastic collar-tray to catch falling seeds and droppings.

Almost all cage birds are seedeaters, therefore feeding is a simple matter. Canaries are fond of canary seed (a mixture of grass seeds) mixed with red rape seed. Both are available in packages from pet stores. The birds also like hemp and linseed, but these are rather oily, therefore fattening, so should only be dispensed as treats.

Canaries also should be given fresh greens and fruits. The greens can be almost anything in that category—lettuce, cabbage, or Brussels sprouts, for instance, while the fruit can be small pieces of apple, pear or grape. The greens and fruit vary the diet, and, unlike humans, the birds will stop eating when their appetites are satisfied.

Budgerigars like canary seed too, but this should be mixed

with millet seed. These miniature parrots should also be given their fill of greens and fruit. They can be fed small pieces of bread moistened with milk or water, as special treats.

Both canaries and budgerigars consume large amounts of grit—which is ground stone, sand, shell and bone—and that particular container should always be filled. This substance not only helps the birds digest their food, but supplies essential minerals to their diet.

Parrots and parakeets are also seedeaters. However, along with their canary seed, millet, hemp and linseed, they should be given their special delight—sunflower seeds. These larger cage birds will also require greens and fruit, plus another delight—nuts, not salted ones, but those in shells, such as almonds, pecans and others.

Although most cage birds will thrive on the diets just described, some of them, like humans, will like a few items on their menus better than others. Only by experience gained from trial and error will owners discover whether their birds like one seed over another, or if this or that fruit or green will be left in the feeding bowl.

But that is some of the enjoyment of having birds as pets. They belong to you, and are your responsibility. If you make sure that they are fed and cared for properly, they will repay your attention many times over. One final suggestion about feeding: be certain that at the end of each day all food—seeds, greens and fruit—that has not been consumed is taken away. Begin each day with a fresh and clean supply.

Having one bird is enjoyable. Having two is even better. And if you do have two, especially canaries or budgies, you can, with little effort and expense, have a most thrilling experience. You can help them to have offspring.

If you do have two small bird pets, make sure the cage is large enough to accommodate both without crowding. Also, if breeding is your intent, make sure they are of different sex. (Pet store proprietors have many amusing tales of clients who have bought two birds without establishing that most important fact.)

Let them become acquainted before expecting results. If one is male and the other female, nature will, eventually, have its way. But you can, with delicacy and patience, be of some help.

Birds usually enter their breeding season during March and April. That's when the male appears to strut and sings almost constantly, while the female appears to become even more demure.

Beginning the first of March place a nesting box in the cage. Various types of these can be obtained from a pet store, but it can be a simple thing—a saucer, or even a small aluminium container that held frozen food. Put a layer of sand or sawdust on the bottom, and beside it a small amount of dry grass and moss, or a mixture of both, obtained from a garden or a drive in the country. Also, if there is a poultry store in your area ask for a small amount of feathers to line the nest.

The birds will do the nest-building, and will appreciate being let alone while construction is in progress. If the twosome is compatible, eggs will appear in about two weeks. Carefully remove the first three and store them in a box lined with something soft: absorbent cotton is excellent. Return them to the nest after the fourth egg has been laid. This will assure that all the eggs will hatch at about the same time.

The incubation period is from 12 to 14 days. As the chicks appear more food should be placed in the cage, with the addition of pieces of bread moistened with milk or water. The parents will take charge of the feeding. But make sure all uneaten food is removed at the end of the day.

The young birds should be removed from the care of their

parents as soon as they are able to feed themselves. When placed in their own cage these youngsters can then be given the usual seed, greens and fruit. And you'll have a family, either to disperse among friends, or to sell to the pet store.

Breeding parrots and parakeets is a trifle more difficult, and is usually done, not in cages, but in larger aviaries, where the birds can fly for short distances. These are often constructed in attics and garages. The procedure is the same: nest-building materials made available, and privacy.

And so, whether you have one bird pet or two, or a roomful of feathered beauties who fill the days with brightness and song, you will be in a world of amazing interest and satisfaction.

Index